Mineral Belts of Western Sierra County New Mexico

by US Dept. of Interior

with an introduction by Kerby Jackson

Introduction

It has been decades since the US Dept. of Interior released their important publication "Mineral Belts of Western Sierra County, New Mexico". First released in 1979, this important volume has been out of print and has been unavailable to the mining community since those days, with the exception of expensive original collector's copies and poorly produced digital editions.

It has often been said that "*gold is where you find it*", but even beginning prospectors understand that their chances for finding something of value in the earth or in the streams of the Golden West are dramatically increased by going back to those places where gold and other minerals were once mined by our forerunners. Despite this, much of the contemporary information on local mining history that is currently available is mostly a result of mere local folklore and persistent rumors of major strikes, the details and facts of which, have long been distorted. Long gone are the old timers and with them, the days of first hand knowledge of the mines of the area and how they operated. Also long gone are most of their notes, their assay reports, their mine maps and personal scrapbooks, along with most of the surveys and reports that were performed for them by private and government geologists. Even published books such as this one are often retired to the local landfill or backyard burn pile by the descendents of those old timers and disappear at an alarming rate. Despite the fact that we live in the so-called "Information Age" where information is supposedly only the push of a button on a keyboard away, true insight into mining properties remains illusive and hard to come by, even to those of us who seek out this sort of information as if our lives depend upon it. Without this type of information readily available to the average independent miner, there is little hope that our metal mining industry will ever recover.

This important volume and others like it, are being presented in their entirety again, in the hope that the average prospector will no longer stumble through the overgrown hills and the tailing strewn creeks without being well informed enough to have a chance to succeed at his ventures.

Kerby Jackson
Josephine County, Oregon
August 2015

CONTENTS

FIGURES

Mineral Belts in Western Sierra County, New Mexico, Suggested by Mining Districts, Geology, and Geochemical Anomalies

By Tom G. Lovering *and* Allen V. Heyl

Abstract

More than 80 percent of the total mineral production of Sierra County has come from mining districts that lie within two relatively narrow northerly trending belts, in the western part of the county; a western belt, which lies along the eastern flank of the Black Range-Mimbres Mountains, and an eastern belt which includes the Sierra Cuchillo at the northern end, and extends southerly through the Salado Mountains, Animas Hills, Apache Hill, Town Mountain, and terminates about a mile north of Greg Hills near the southern boundary of Sierra County. The western belt, including the old mining districts of Phillipsburg, Grafton, Chloride, Hermosa, North Percha Creek, Kingston, South Percha Creek, and Tierra Blanca, has produced approximately $20 million worth of ore mostly from a few large mines in the Chloride, Hermosa, and Kingston mining districts. The eastern belt, including the Iron Mountain, Cuchillo Negro, Chise, Salado Mountains, Hillsboro, Lake Valley, and Macho districts and mineralized areas, has produced about $15 million worth of ore, chiefly from mines and placers in the Hillsboro district, and mines in the southern part of the Lake Valley district.

Geological and mineralogical information on the known ore deposits in these two mineral belts is combined with information from geochemical studies derived from published maps and reports, and unpublished information from the maps, field notes, and files of the authors.

This study has resulted in the identification of several new areas in both mineral belts in which geology, geochemistry, or both, are favorable for finding undiscovered mineral deposits. These areas include the western part of the Chloride district, east of the Silver Monument mine; the northern part of the main Chloride district, south of Chloride Creek; the southern part of the Hermosa district, south of the American Flag and Flagstaff mines; the Kingston district, an area south of Kingston; and the southern part of the Tierra Blanca district (eastern part), east of the mouth of Cottonwood Creek in the western belt. Favorable areas for exploration in the eastern belt include Jaralosa Mountain, the northern part of the Cuchillo Negro district north of Red Hill, the southeastern part of the Cuchillo Negro district east of HOK Ranch, the Salado Mountains area, the southern part of the Hillsboro district south of Copper Flat, the area just south of the Lake Valley district north of Town Mountain, and the old Macho district.

INTRODUCTION

This report covers the part of Sierra County, New Mexico, that lies west of the Rio Grande. It, thus, excludes the deposits in the Caballo Mountains of lead, vanadium, fluorspar, and gold, and also the old Spanish gold mines near the northern end of the Cristobal Mountains east of the Rio Grande. Approximately 80 percent of the total mineral production of Sierra County has come from mining districts which fall within two relatively narrow northerly trending belts, a western belt which covers part of the eastern flank of the Black Range-Mimbres Mountains from the northern border of the county to about 12 miles north of the southern boundary, and an eastern belt which includes the western part of the Sierra Cuchillo, the Salado Mountains, Animas Hills, Apache Hill, Town Mountain, and terminates about a mile north of Greg Hills near the southern border of the county. The western belt includes the old mining districts of Phillipsburg, Grafton, Chloride, Hermosa, Kingston, and Tierra Blanca. The eastern belt includes Iron Mountain, Cuchillo Negro, Chise, Salado Mountains, Hillsboro, Lake Valley, and Macho mining districts and mineralized areas (fig. 1).

A. V. Heyl has mapped the geology of the Iron Mountain, Chise, and Priest Tank quadrangles, and portions of the Winston and Jaralosa Mountain quadrangles. T. G. Lovering has studied jasperoid samples from several localities in the area covered by this report, particularly the Cuchillo Negro, Hermosa, Salado, Hillsboro, and Lake Valley districts, as well as

Manuscript approved for publication, September 6, 1988.

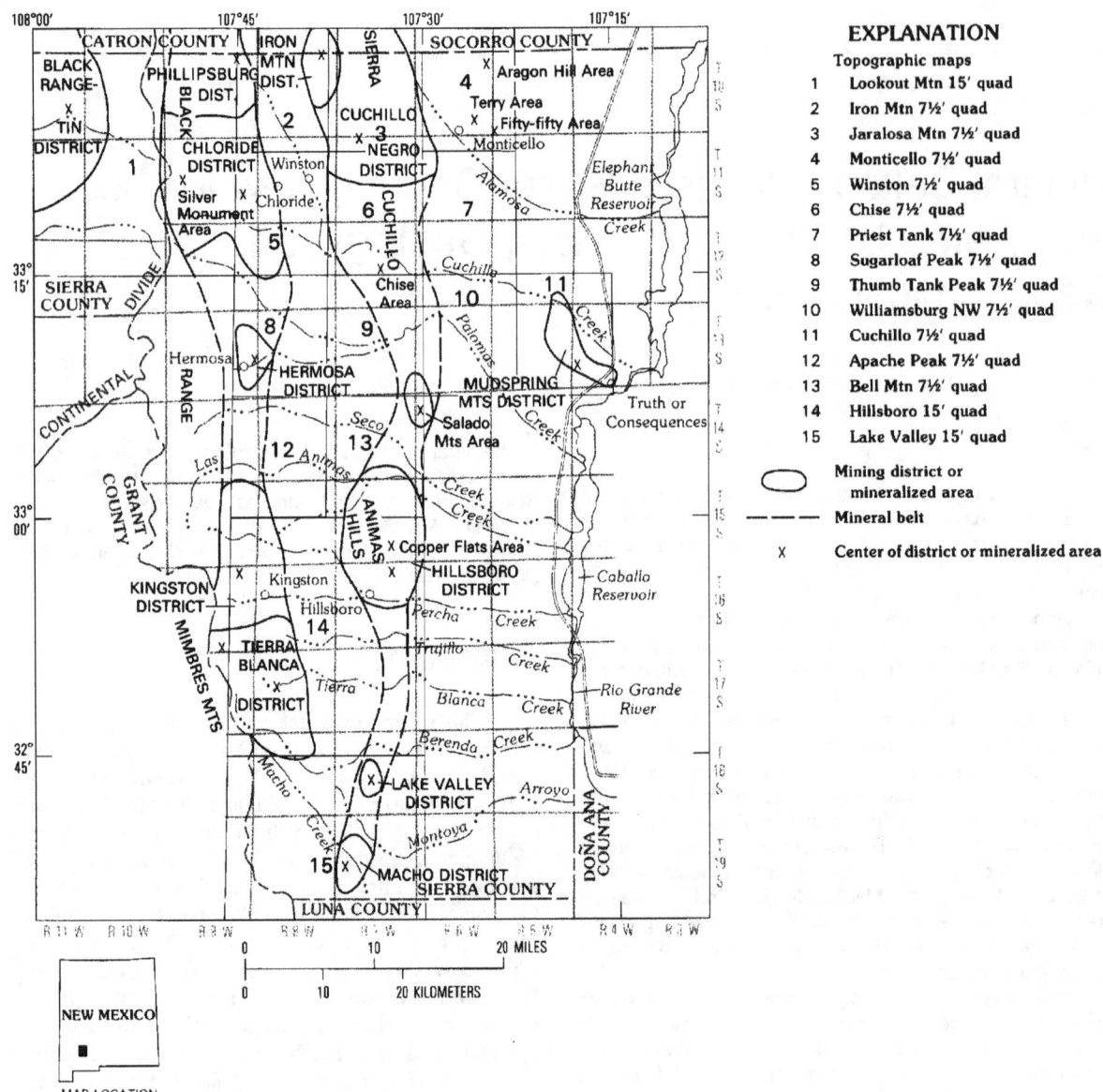

Figure 1. Sketch map of western Sierra County showing general location of mining districts and mineralized areas.

magnetic concentrates and calcite from a few localities. Our colleague, C. H. Maxwell, has mapped the Sugarloaf Peak, and portions of the Winston, Lookout Mountain, Thumb Tank Peak, Williamsburg NW, Cuchillo, Apache Peak, and Bell Mountain quadrangles; and both Heyl and Maxwell have visited and sampled all the districts and mineralized areas discussed in this report.

Our investigations have been preceded by those of many other workers, whose published contributions include the following reports and maps: Clark (1894);

Harley (1934); Hill (1946); Jahns (1944, 1955a, b); Jicha (1954a); Kuelmer (1955); Lindgren, Graton, and Gordon (1910); McAnulty (1978); Maxwell and Heyl (1980); Creasey and Granger (1953); Dane and Bachman (1961); Hedlund (1977a, b); and Jicha (1954b). The report also incorporates geochemical anomaly data on heavy-mineral concentrates from stream-sediment samples, contained in published maps by Alminas and others (1975a, b, c), by Watts, Alminas, and Kraxberger (1978), and by Watts and others (1978a, b, c), together with published information on jasperoid samples by Young and Lovering (1966) and Lovering (1972).

The geologic features discussed in this report reflect recent, careful field studies by Heyl and Maxwell, whose interpretation of fault locations and bedrock geology in some areas differs from that depicted on older published geologic maps, which were basically reconnaissance maps rather than detailed field maps, specifically those by Alminas and others (1975a, b, c) covering the region north of 33° N. lat., the map by Hedlund (1977a) covering the Hillsboro 15' quadrangle, and the map by Jicha (1954a) covering the Lake Valley 15' quadrangle.

This report includes brief summaries of each mineral belt, which are followed by more detailed discussions of the mineralization in each belt, starting at the north end and working south, and covering the location and size, production, geologic setting, and geochemical information on each mining district or mineralized area, with brief general discussions of the geology and geochemical anomalies in the unmineralized or slightly mineralized gaps that separate them. This is followed by a section containing brief discussions of deposits in western Sierra County that lie outside the mineral belts, and a section on suggestions for prospecting, containing geological and geochemical evidence for undiscovered mineral deposits in various parts of the mineral belts.

The locations of geologic features and geochemical samples referred to in the text are given with reference either to Government Land Office grid coordinates (township, range, section) or to local topographic and cultural features that are shown on the 7½-minute or 15-minute topographic maps covering the areas. It is not practical to reproduce these large-scale topographic maps for this report, but they are all shown and listed in figure 1. It is hoped that combining geological and geochemical information from many sources in this report will provide both a more comprehensive view of the known deposits in the mineral belts and useful information on promising areas for further exploration.

ACKNOWLEDGMENTS

The authors are deeply indebted to C. H. Maxwell, who not only provided the index map and base maps for mining districts in the northern part of the western mineral belt, and has critically reviewed this portion of the manuscript, but also has provided many jasperoid samples and much valuable information on geology and mineralization in the entire region covered by this report. Mr. Maxwell has contributed so much that he deserves, but will not accept, co-authorship of the report. Other colleagues who have provided geochemical samples and information on mining districts and mineralized areas include W. R. Griffitts, K. C. Watts, J. A. Hedal, and M. R. Oakman. Teresa Feder drafted the original map illustrations of the mining districts.

GENERAL SUMMARY OF THE WESTERN MINERAL BELT

The western mineral belt extends southerly along the eastern flank of the Black Range and Mimbres Mountains for approximately 45 miles, from the northern end of the Phillipsburg district a few miles north of the Sierra-Catron County line to the southern end of the Tierra Blanca district (eastern part) about 12 miles north of the Sierra-Luna county line. It includes, from north to south, the Phillipsburg, Grafton, Chloride, Hermosa, Kingston, and Tierra Blanca mining districts (fig. 1). The belt is discontinuous with several less mineralized, unmined gaps; the largest gap is about 7 miles long between the Hermosa and Kingston (North Percha Creek area) districts. There also appears to be an offset of about 3½ miles to the east between the western and eastern parts of the Tierra Blanca district near the southern end of the belt.

The total value of the metal produced from the districts in this belt can only be estimated in a general way, as early production records of many of the old mining camps are incomplete, but through 1983 it probably amounts to about $20 million. Nearly half of this amount has come from the St. Cloud mine in the Chloride district since it reopened in 1980. Other major mines which have produced more than $1 million worth of ore include the Pelican mine in the Hermosa district and the Andy Johnson and Lady Franklin mines in the Kingston district.

All the mining districts in the western belt have been silver producers. Those at the northern end (Phillipsburg, Grafton, and Chloride) have also been important gold producers, but the ratio of gold to silver in the ore declines southward from Phillipsburg, where it is about 1:50, to Chloride (1:100), to Hermosa, where it is about 1:1000. The only district in the southern part of the belt which produced much gold is the Tierra Blanca district. Lead has been a major mineral commodity in the southern part of the Chloride district and in all the districts to the south of it. Copper is important in the Grafton district and in the central and western parts of the Chloride district, as well as in some of the mines of the Kingston and South Percha Creek areas. Although zinc is commonly associated with copper and lead in ore from the southern part of the Chloride district, the Hermosa district, and the Kingston districts, very little zinc has been produced because of adverse market conditions during the period when the mines were operating. Some tungsten has been produced in the North Percha Creek area, and some manganese ore was mined in the main Kingston district.

Host rock for the ore deposits consists largely of older Tertiary andesitic volcanic rocks in the Phillipsburg and Grafton districts at the northern end of the belt. These rocks also contain the largest and richest deposits in the Chloride district, although some ore has also been produced from mines in Pennsylvanian Magdalena Group limestone. In the Hermosa district most of the ore has come from carbonate beds in the Silurian Fusselman Dolomite and Ordovician Montoya Group, although some has also come from limestones in the Pennsylvanian Magdalena Group and the Lower Mississippian Lake Valley Limestone. Precambrian granitic rocks are hosts for most of the vein deposits in the North Percha Creek area. At Kingston most of the ore is in carbonate rock of the Silurian Fusselman Dolomite below the Upper Devonian Percha Shale. Lower Ordovician El Paso Limestone (or Group) contains the ore bodies of the South Percha Creek area. In the western part of the Tierra Blanca district, rich gold ore was mined from the Upper Cambrian Bliss Sandstone, with lower grade silver-lead deposits in the overlying Ordovician El Paso Limestone. In the eastern part of the Tierra Blanca district, as at Kingston, ore deposits are in the carbonate rocks of the Fusselman Dolomite and Montoya Group beneath the Percha Shale.

Ore deposits are locally associated with felsic dikes in the northern part of the mineral belt in the Phillipsburg district, and in the south-central and southern part in the South Percha Creek area and western and eastern Tierra Blanca district. At Kingston many of the felsic dikes are close to the border of an altered monzonite stock.

Ore deposits exhibit a strong structural control throughout the western mineral belt (pl. 1). The veins trend mostly northerly and northeasterly in the Phillipsburg and Grafton districts, and northwesterly in the Chloride district. In the Silver Monument district, small, rich silver-copper veins trend northeasterly, northwesterly, and easterly. At Hermosa, strong northerly and northwesterly trending mineralized faults are crossed by smaller easterly trending faults in the central part of the district (Maxwell and Heyl, 1980; Jahns, 1955a). In the North Percha Creek area, larger fault veins trend northerly, but smaller easterly trending faults are also mineralized. The main northerly trend continues through the eastern part of the Kingston district, but in the western part, near the intrusive stock, the main trend is northwesterly, with some mineralization related to northeasterly trending faults as well. In the South Percha Creek area, ore is localized along a strong northwesterly trending fault. In the western part of the Tierra Blanca district, small northerly

and northeasterly trending faults control mineralization. In the eastern part of this district, ore deposits occur in close proximity to a strong northerly trending fault zone, which appears to be cut and displaced westward in the northern part, along a big easterly trending fault zone.

Wall-rock alteration in the western mineral belt varies with the type of host rock. Where mineralized faults and fissure veins cut Tertiary andesite, the wall rock is propylitically altered in an outer zone extending as much as 30 ft outward from the fault or vein; within this zone is a much narrower casing of silicified and sericitized rock, which commonly also contains talc and pyrite. Mineralized veins in carbonate rocks generally have an outer envelope of hydrothermal dolomite with disseminated pyrite and an inner casing of silicified rock (jasperoid) in some deposits. Most of the ore bodies in carbonate rocks are pod-shaped bodies along faults, and at fault intersections; these ore bodies are encased in envelopes of hydrothermal talc and sericite. Unmineralized jasperoid also forms extensive blankets at the base of the Percha Shale in the western part of the Tierra Blanca and in the Kingston districts; it is abundant above the Percha Shale at Hermosa. Where veins cut Precambrian granitic rocks in the North Percha Creek area, the wall rock is silicified in an outer zone and has a narrower zone of pyritic and sericitic alteration adjacent to the vein wall.

All the mineral deposits in the western belt are considered to be of middle or late Tertiary age, younger than the older andesitic and most of the felsic volcanic rocks. The common association of mineralized structures with felsic dikes in the Phillipsburg, southern Kingston, and Tierra Blanca districts suggests, but does not prove, a genetic relationship between the deposits of these districts and these intrusive rocks of probable Oligocene age.

Geochemical samples collected in the western mineral belt include a few calcite vein samples near the northern end in the Phillipsburg district, a few jasperoid samples including some in andesite from the same area, two more from the Chloride district, and a rather extensive suite from the Hermosa district. Panned heavy-mineral-concentrate samples have been collected from small streams and dry washes throughout the length of the western belt. However, the number of such samples collected ranges from several per square mile in the Hermosa and Kingston districts, down to one sample in several square miles in some of the apparently unmineralized gaps in the belt, such as the 7-mile gap between the North Percha Creek area and the Hermosa district. The calcite dike samples from the northern end of the belt, though few in number, exhibit base- and precious-metal anomalies related to known mineral deposits, suggesting that this sample medium is worthy of further investigation. Jasperoid samples from the same area are mostly barren of anomalous concentrations of ore and ore-related elements. However, such anomalies are present in two jasperoid samples from the Chloride district, and in most of those collected from the Hermosa district. Metal anomalies in nonmagnetic heavy-mineral concentrates discussed in this report are limited to samples that contain one or more of the following metal concentrations: [1]$Zn \geq 10,000$ ppm, $Pb \geq 2,000$ ppm, $Cu \geq 1,000$ ppm, $Sn \geq 1,000$ ppm, $W \geq 1,000$ ppm, $Ag \geq 100$ ppm, $Bi \geq 100$ ppm, $Mo > 70$ ppm; and also fluorite visible in the concentrate. These concentrations correspond to the highest values given on the regional geochemical maps (Alminas and others, 1975a, b, c; Watts, Alminas, and Kraxberger, 1978; Watts and others, 1978a, b, c). These maps also show lower concentration anomalies for each element, ranging from 1/7th to 1/10th of the concentrations listed above. Lower level anomalous concentrations have been omitted from the text discussion and the district maps, partly to avoid clutter on the maps, and partly to reduce the size of this report, as the large numbers of such moderately anomalous samples shown on the source maps would require a considerable increase in the volume of this report to adequately discuss. Where base- and precious-metal deposits are in veins cutting nonreactive host rocks, such as andesite or granite, silver anomalies seem to provide the longest downdrainage anomaly trains, extending as much as 2 miles below the source. However, in areas where these deposits are hosted by carbonate rocks, silver anomalies rarely extend more than a quarter of a mile below their source, and the longest anomaly trains are generally provided by lead and molybdenum. Compound metal anomalies involving copper and (or) zinc in addition to lead, silver, and molybdenum generally indicate close proximity to an ore deposit in the upstream drainage. Anomalies for tin and bismuth are few and scattered, generally occurring with other metal anomalies a short distance downdrainage from a mine. Tungsten anomalies are present in the vicinity of iron-manganese-tungsten fissure vein deposits in the North Percha Creek area; however, notable tungsten-molybdenum anomalies commonly occur together in many stream-sediment concentrate samples from Kingston, the South Percha Creek area, and the Tierra Blanca district for a mile or more downdrainage from the nearest mine. This association is somewhat surprising since neither metal has been produced from the mines of these districts. Visible fluorite in pan concentrates of heavy minerals commonly occurs in samples devoid of metal anomalies in widely separated localities from the central and southern part of the belt. Fluorite is abundant in samples from the largely volcanic terrain west of the

southern end of the belt. There is also an area with abundant tin anomalies in streams draining the southern Mimbres Mountains south of the western part of the Tierra Blanca district.

Metal anomalies in heavy-mineral concentrates from the drainages devoid of mines or prospects in the upstream drainage basins are present at many localities throughout the belt. Such anomalies may be attributable to the presence of undeveloped ore deposits. The most significant of these are (1) near the southern end of the Chloride district; (2) east and south of the main Hermosa district; (3) east of the North Percha Creek area; (4) south of the Kingston district; (5) west of the northern end of the eastern part of the Tierra Blanca district; and (6) at the southern end of the Tierra Blanca district. These anomalies are discussed in the following section of this report.

MINERALIZATION IN THE WESTERN BELT

Phillipsburg and Grafton Districts

Location

The northernmost well-known mineralized area in the western mineral belt covers about 15 square miles in the northern part of the Black Range in Catron and Sierra Counties. It extends from the Silver Creek drainage in Catron County southward approximately to the divide between Turkey Creek and Bear Creek in Sierra County, a distance of about 6 miles, with a maximum width of about 3 miles, and includes the old mining camps of Phillipsburg (fig. 2) in the northern part of the area, and Grafton in the southern part.

Production

This area produced approximately $1 million worth of ore prior to 1980, mostly gold with some silver and a little copper. The main producing mines in the area were the Occidental, Minnehaha, Great Republic, Keystone, and Gold Bug in the northern (Phillipsburg) part of the area, and the Emporia, Ivanhoe, and Alaska mines in the southern (Grafton) part of it (Harley, 1934, p. 78–81). All these produced principally gold ore with some silver; mines around Grafton also produced copper (Harley, 1934, p. 78–81). Several of the larger mines in the district were reopened in the early 1980's, and the Occidental,

---
[1]Note: ppm indicates parts per million.

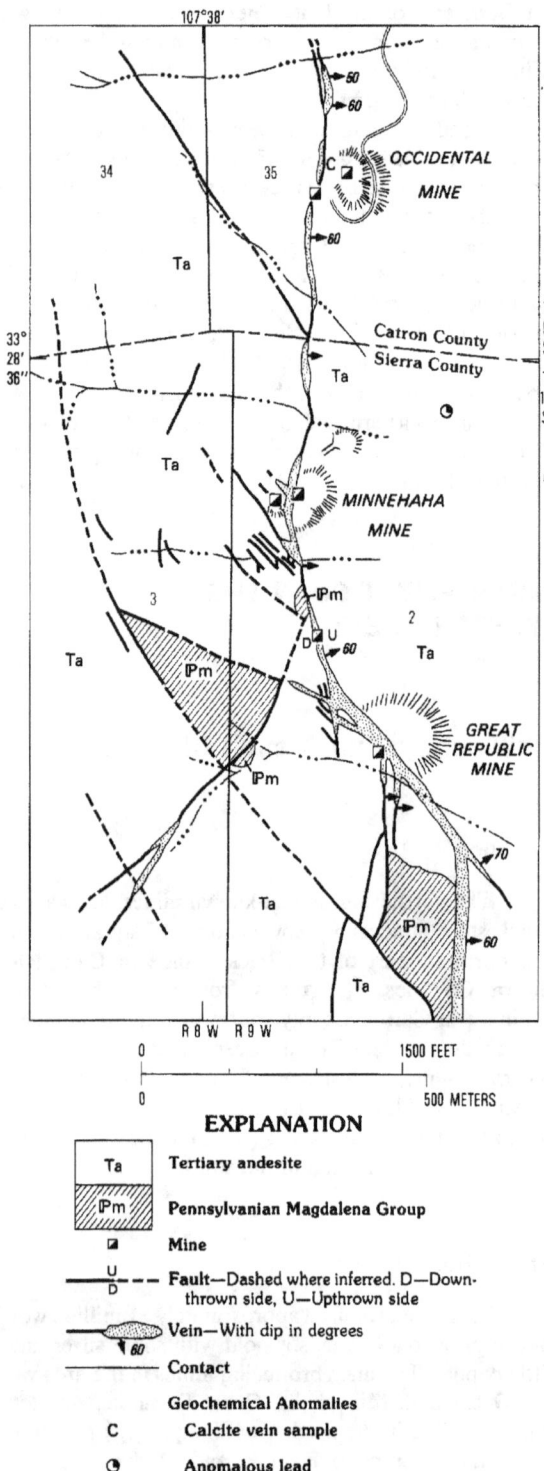

34 35

C ▣ OCCIDENTAL MINE

Ta

Catron County
Sierra County

T 9 S

T 10 S

Ta

Ta

MINNEHAHA MINE

3

Pm

Ta

Pm

D U 60

2

Ta

Pm

GREAT REPUBLIC MINE

Pm

Ta

70

Ta

Pm

60

Ta

R 8 W   R 9 W

0 — 1500 FEET

0 — 500 METERS

EXPLANATION

| Ta | Tertiary andesite |

| Pm | Pennsylvanian Magdalena Group |

▣ Mine

U/D – – – Fault—Dashed where inferred. D—Down-thrown side, U—Upthrown side

Vein—With dip in degrees
60

Contact

Geochemical Anomalies

C   Calcite vein sample

☾   Anomalous lead

Figure 2. Sketch map of central Phillipsburg district showing approximate location of geologic features, principal mines, and anomalous geochemical sample localities (modified from Maxwell and Heyl, 1980).

Minnehaha, and Great Republic were still major producers of siliceous gold-silver flux ore in 1987. The Emporia and Ivanhoe mines at Grafton were producing rich gold ore in 1985–87.

Geologic Setting

The area is largely covered by highly altered andesite flows of Late Cretaceous or early Tertiary age; less-altered andesite of Oligocene age is also present. Exotic gravity slide blocks of Paleozoic carbonate rock (dominantly Pennsylvanian Magdalena Group) overlie propylitized andesite, which may be early Tertiary in age. The early Tertiary andesite is overlain by a younger andesites. In addition, small horses of deformed carbonate rocks are locally present along strong fault veins cutting the andesites. A nearly circular mass of highly altered rhyolite breccia, about half a mile in diameter, occupies the northeastern part of the area just north of the junction of Wildhorse Canyon and Sheep Canyon. Ash flow tuffs and rhyolite flows and dikes related to this center extend southward about 1½ miles as far as Poverty Creek.

The ore occurs along strong northerly, northwesterly, and northeasterly trending quartz veins, some of which can be traced continuously for more than a mile and are as much as 8 ft wide in places. The veins dip very steeply, and many are nearly vertical. They are encased in a broad zone of propylitically altered rock enclosing a much narrower casing of silicified andesite adjacent to the veinwall. Locally, in the northern part of the area, rhyolite dikes lie parallel and adjacent to the veins. Ore bodies are generally in the form of small masses of breccia on either the hanging wall or footwall of the vein, in which the breccia fragments are partly replaced and partly cemented by ore, or in thin dark bands near the vein walls cutting white quartz beneath amethyst. In the northern part of the area, north of Poverty Creek, the ore consists dominantly of very fine grained gold particles in the dark bands in quartz or amethyst veins which locally contain supergene copper and silver minerals and have sparse sulfides. Towards the southern part of the area the copper and silver content of the ore increases and sulfides become more abundant (Maxwell and Heyl, 1980).

In addition to quartz, calcite is a locally important gangue mineral, but where it is abundant, metal values are low in the quartz. Amethyst is an important indicator of gold ore in the area. Ore minerals in the northern veins are free gold, auriferous pyrite, and locally cerargyrite and bromargyrite. Sulfide ore minerals include acanthite, argentite, tetrahedrite, bornite, and chalcopyrite, which together with their alteration products, are most abundant in mines like the Ivanhoe and Emporia near the southern end of the area.

Geochemical Information

Rock samples.—Jasperoid samples of silicified wall rock adjacent to the veins were collected from outcrops along a strong northeast-trending vein that grazes the northwest edge of the rhyolite breccia plug in the northern part of the area, and also from mine dumps near the Occidental, the Minnehaha, and Great Republic mines which are located on strong northerly trending quartz veins that cut the andesite (fig. 2). One outcrop sample of aphanitic yellowish gray silicified rhyolite next to the plug contains as much as 0.5% sodium, 7% potassium, 150 ppm cerium, 70 ppm lanthanum, 70 ppm neodymium, and 300 ppm zirconium. Jasperoid dump samples from the gold mines are variable in texture, generally light gray in color with accessory pyrite, calcite, and sericite. Most of the jasperoid shows anomalously high vanadium (30–70 ppm), and some samples also contain higher than normal concentrations of lanthanum (20 ppm), and molybdenum (70 ppm).

A sample of vein calcite from the dump of the Occidental mine, in the northern part of the area, contains anomalously high concentrations of 0.4 ppm gold, 43 ppm silver, 7 ppm arsenic, 0.4 ppm tellurium, 20 ppm copper, and 12 ppm lead. One from the dump of the Gold Bug mine in sec. 10, T. 10 S., R. 9 W., contains anomalous 14 ppm arsenic and 0.2 ppm tellurium. An outcrop sample of calcite from a vein crossing Poverty Creek near the southeast corner of sec. 2 (T. 10 S., R. 9 W.) shows detectable amounts of 0.004 ppm gold and 10 ppm cerium.

Stream-sediment concentrates.—Panned heavy-mineral concentrates collected at two localities close to the rhyolite breccia plug each contain more than 1,000 ppm tin. One such sample from a stream within the rhyolite contains more than 70 ppm molybdenum. One sample from a stream draining the northeasterly trending vein system north of the rhyolite mass, near the north boundary of sec. 26, T. 9 S., R. 9 W., contains more than 100 ppm silver, as do also three samples from Poverty Creek, one at its confluence with the south-draining tributary that runs past the Minnehaha and Great Republic mines, one about 500 ft downstream from this junction, and one about a mile downstream. One sample from near the head of this south-draining tributary just south of the county line contains more than 2,000 ppm lead (Alminas and others, 1975a, b, c).

Summary and evaluation of geochemical samples.—Samples of vein calcite appear to provide better ore-related geochemical anomalies than do samples of jasperoid from this area. Most of the jasperoid appears to reflect early, pre-ore alteration. Silver in the pan-concentrate samples seems to give the longest downstream anomaly trains. There are no geochemical samples available from the southern part of the area.

Chloride District

Location

The mineralized area which centers on the old Chloride mining district (fig. 3) is larger and more complex than the Phillipsburg-Grafton area to the north. It extends from just south of Bear Creek southward to Monument Creek, about 9 miles. In its central part it extends westward from Chloride about 8 miles to Lookout Mountain at the head of Chloride Creek. The total size of this mineralized area is about 40 square miles.

Production

Mines in this area produced about $500,000 worth of mixed gold, silver, copper, and lead ore with some zinc between 1870 and 1930. The district was largely dormant until about 1970 when the increase in price of the precious metals spurred renewed activity and several mines were reopened. The chief mines in the northern part of the area between Bear Creek and Mineral Creek are the Black Hawk, Bellboy, Dreadnaught, Paymaster, Gold Hill, and Readjuster. In the central part of the district close to Chloride Creek, are the Wall Street, Nana, and Hoosier mines to the east near Chloride. The Silver Monument mine is located to the west near the head of Chloride Creek. Farther south between Hagins Peak and Chloride Creek are the St. Cloud and U.S. Treasury mines, both of which were active in 1985; the Colossal mine, also active, is located on the north side of South Fork Creek, and in the southern part of the district between South Fork and Monument Creek, are the Midnight mine, Pye Lode, and Bald Eagle mines (Harley, 1934, p. 81–90). Two mines in the district, the St. Cloud and the U.S. Treasury, since they were reopened in 1980, have together produced about $10 million worth of silver, gold, and copper ore by 1985. They were still in operation, as well as the Colossal, in 1987. Some of the St. Cloud ores are very rich in gold, copper, and silver.

Geologic Setting

This area is considerably more complex, geologically, than the Phillipsburg-Grafton area to the north. Although the most abundant rock in the area is altered early Tertiary andesite and dacite, this rock is locally capped by thick flows of younger Tertiary andesite and rhyolite. Exotic blocks of Magdalena Group limestone, locally, with the overlying Lower Permian Abo Formation, which rest on early Tertiary andesite, are much larger and more numerous in the Chloride area than they are to the north; some blocks cover areas of a square mile or more. Although the older andesites are the most favorable host rock for ore deposits, mines have been developed in all three types (Maxwell and Heyl, 1980). The Readjuster and Bellboy mines in the northern part of the area are in latite and rhyolite tuff. The Dreadnaught and Hoosier mines are in limestone of the Magdalena Group. The U.S. Treasury, Bald Eagle, Wall Street, Nana, and Silver Monument mines are all in older andesite. The old upper workings of the St. Cloud mine are in limestone, but the large, rich new ore bodies are in older andesite beneath and on the southwest side of the limestone block. Most of the veins in the northern part of the area, north of Mineral Creek, trend northerly or north-northeasterly, as they do in the Phillipsburg-Grafton area. In the central-eastern part, between Mineral and South Fork Creeks, the veins trend northwest, or west-northwest, and some are cut by large north-northeast and north-trending barren faults (fig. 3). In the vicinity of the Silver Monument mine at the head of Chloride Creek, in the central-western part of the area, veins trend north-northeast, west-northwest, and northwest. In the southeastern part of the area between South Fork and Monument Creeks, veins trend northerly or north-northwesterly. Veins tend to widen where they pass from limestone into andesite; however, the richest parts of the veins opened recently have andesite on one wall and limestone on the other

In andesite, veins generally have an alteration halo of intense propylitic alteration extending several feet into the adjacent rock with a much narrower zone of silicic and pyritic alteration, locally accompanied by talc adjacent to the vein walls. In limestone, vein walls are commonly silicified and pyritized for a few feet outward from the veins in the vicinity of ore bodies. In a few places, as at the Dreadnaught mine, wall-rock alteration adjacent to veins in limestone consists of argillic alteration and talc. Alteration along veins cutting rhyolite consists of narrow zones of argillization and silicification.

The nature of the wall rock appears to influence the type of ore in this area, although this difference may also reflect primary hydrothermal zoning. Ore bodies in veins cutting rhyolite produce mainly silver with some gold, and locally copper. Those in andesite are chiefly copper-silver and gold producers. Those in limestone are producers of mixed copper-lead-zinc with some silver and very little gold. The Wall Street mine on the Apache vein system in andesite on the north side of Chloride Creek produced mostly free gold and auriferous pyrite in a quartz gangue; whereas, the Hoosier mine a half a mile to the southeast on the same vein in Magdalena Group produced oxidized silver-lead ore in a gangue of calcite, limestone breccia, and talc.

The assemblage of ore and gangue minerals varies from mine to mine, and even from ore body to ore body in the same mine as a function of host rock and depth of

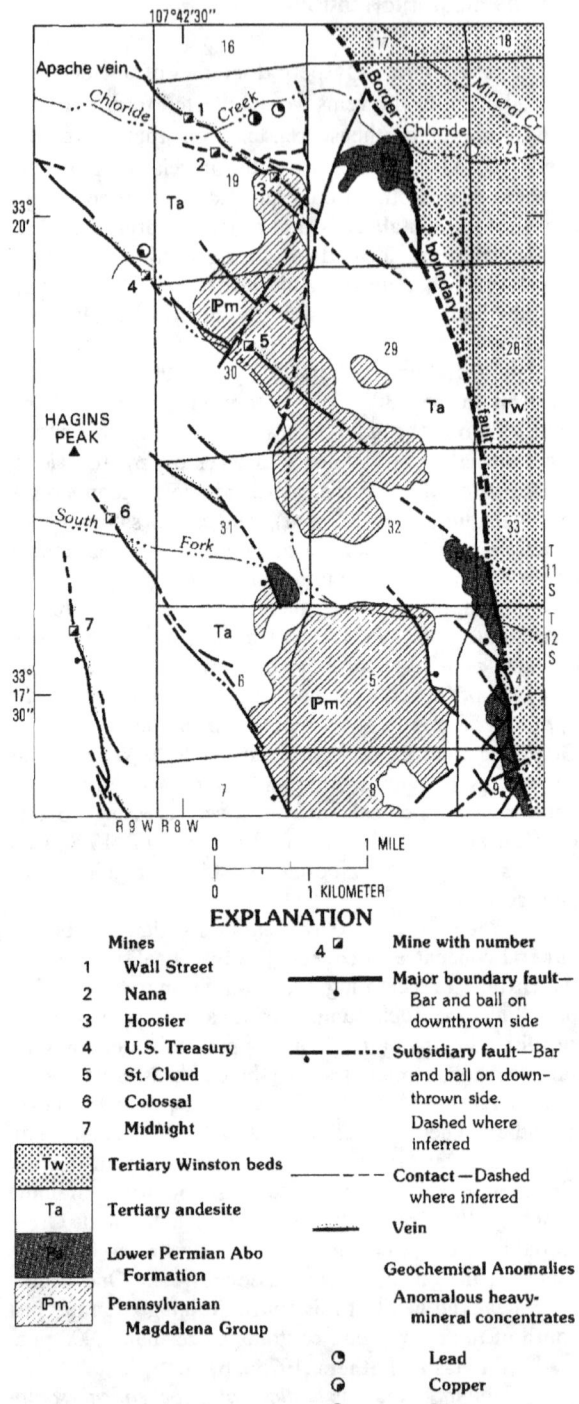

Figure 3. Sketch map of main Chloride district showing generalized location of geology and structural features (modified from Maxwell and Heyl, 1976), principal mines, and anomalous geochemical sample localities.

emplacement. In the upper workings of the St. Cloud mine, 10-ft- to 15-ft-wide veins in Magdalena Group contained sparse galena, sphalerite, chalcopyrite, a little argentite, and little gold. In the upper part of deeper new workings, veins up to 35 ft wide and breccia zones in andesite contain large, rich ore bodies of bornite, silver, electrum, covellite, stromeyerite, chalcopyrite, and argentite with a little galena and sphalerite in a gangue of quartz, talc, chlorite, and pyrite; in the lowest workings on St. Cloud ore shoot, ore minerals have changed to galena, sphalerite, and chalcopyrite. This same type of rich copper-silver-gold ore in andesite host rock also characterized the Silver Monument mine at the western end of the district, and may be present beneath the old workings in some other mines in the central and western parts, such as the U.S. Treasury mine. In general, the mines in the southern part of the area produced mostly silver, copper, lead, and zinc with some gold. Those in the central part produced mainly copper and silver with some lead and a little gold, and those at the northern end mostly gold and silver with some copper (Maxwell and Heyl, 1980).

Geochemical Information

Rock samples.—Unfortunately, jasperoid samples are available from only two localities in this area. One is from a small mine south of Bear Creek just east of Cliff Canyon in sec. 34, T. 10 S., R. 9 W., on the northern edge of the area; the other is from an ore bin at the St. Cloud mine in the central part of the area, representing mineralization in the old upper workings in limestone. The jasperoid from the northern locality is from a breccia zone along a strong north-trending vein cutting dacite of the older volcanics. It consists of angular fragments of medium-grained yellowish-gray, olive-gray, and brownish-gray jasperoid in a matrix of slightly coarser dark-gray silica. Both generations contain limonite pseudomorphs after pyrite, and the younger matrix also contains sparse sphalerite altering to hemimorphite. These samples are strongly anomalous in vanadium (200 ppm)—(supergene descloizite and mottramite have been found on old mine dumps in the area). The sample from the ore bin of the old St. Cloud mine consists of dense aphanitic, medium-gray jasperoid fragments cemented by light-gray calcite, with quartz crystals lining vugs. It contains highly anomalous concentrations of manganese (5,000 ppm), lithium (150 ppm), and copper (1,000 ppm), and it is moderately anomalous in silver (30 ppm), lead (500 ppm), and zinc (3,000 ppm).

Stream-sediment concentrates.—A few pan-concentrate samples have been collected from stream sediments in the northeastern and southeastern parts of the area (Alminas and others, 1975a, b, c). One, from the mouth of a south-flowing tributary of Mineral Creek

about half a mile below the Dreadnaught mine (sec. 18, T. 11 S., R. 8 W.), contains more than 2,000 ppm lead and more than 100 ppm silver. Another sample from Chloride Creek, about 2,000 ft below the point where the Apache vein crosses the creek southeast of the Wall Street mine, contains more than 1,000 ppm copper and more than 2,000 ppm lead. Another sample taken about 1,000 ft farther downstream showed a similar anomaly for lead. A sample from a locality immediately below the place where the U.S. Treasury mine vein crosses a small gulch northwest of the mine contained more than 100 ppm silver. Near the southern edge of the area, a sample about 500 ft down the gulch southeast of the southern shaft of the Bald Eagle mine contained more than 1,000 ppm copper, as did another sample about 1,500 ft farther down the gulch, just above its confluence with a south-draining tributary gulch. A sample from this tributary gulch just above the confluence contained more than 1,000 ppm copper and more than 10,000 ppm zinc, probably derived from small deposits in the upstream drainage basin of this gulch. An isolated sample from a locality just north of the junction of southeasterly and southerly trending tributaries of Monument Creek about 2,000 ft east-northeast of Garcia Well was reported to contain more than 100 ppm silver, but a later check sample collected from this locality by C. H. Maxwell failed to duplicate the anomaly.

Summary and evaluation of geochemical samples.—Although the available geochemical samples from this mineralized area are too sparse to provide an accurate basis for evaluation, they do indicate that both jasperoid samples and pan-concentrate samples contain anomalous concentrations of metals related to ore deposits. In some localities there is a later ore stage jasperoid containing ore-sulfide minerals. In the heavy-mineral concentrates, copper and silver appear to provide the best anomalies downstream from areas of mineralization. More detailed geochemical sampling of this mining district, involving both jasperoid and heavy-mineral-concentrate samples, is recommended.

Monument Creek–Willow Creek Gap

There appears to be a gap, or break, in the western mineral belt extending southward from Monument Creek to Willow Creek, a distance of about 3½ miles; the area includes the drainage basins of the North Fork Palomas Creek, Grapevine Canyon, Pole Corral Canyon, and Poverty Canyon, in which there are no mines and few prospects. Geologically, this area, west of the northerly trending range-front fault (pl. 1), consists of a belt of upper Paleozoic sedimentary rocks that is relatively narrow at the north end and widens to the south. These Paleozoic rocks are covered on the western side by early Tertiary volcanic rocks that extend westward to another major north-trending fault which crosses the upper drainage basin of Monument Creek and extends southward between Cobb Mountain and Sugarloaf Peak and across the head of Willow Creek. On the west side of this fault, the rocks are mostly younger Tertiary volcanics. Paleozoic sedimentary rocks east of this fault consist dominantly of Upper Pennsylvanian Madera Limestone of the Magdalena Group, locally overlain by variegated shales with limestone cobbles, the Lower Permian Bursum Formation, which in turn is locally overlain by continental red sandstone and siltstone of the Abo Formation. The older volcanics are mostly dark colored andesite, latite and dacite flows, and pyroclastics. Younger volcanics to the west of the western major fault are mostly light colored rhyolite flows and tuffs. These layered rocks have been intruded by a rhyolite plug that caps Sugarloaf Peak in the southwest part of the area. A diabase intrusive extends northerly for nearly a mile on the upper east flank of Montosa Mountain, where it cuts Magdalena Group; and a small northwesterly trending quartz monzonite stock cuts Magdalena Group in the canyon of North Fork Palomas Creek about half a mile upstream from Romero Tank, near the southeastern edge of the gap.

The eastern northerly trending range-front fault zone extends south across Monument Creek near Monument Spring, continues southward along the northeastern flank of Montosa Mountain, and crosses the ridge about 1,000 ft east of the diabase intrusive and continues south across the North Fork Palomas Creek about 500 ft west of Romero Tank. A strong northeasterly trending fault extends from the east end of the rhyolite plug on Sugarloaf Peak across Poverty Canyon and Grapevine Canyon and across North Fork Palomas Creek about a half a mile south of Dines Ranch, and finally dies out on the south side of Monument Creek about a half a mile upstream from Monument Springs. This fault is downthrown on the northwest side with Tertiary andesite to the north in contact with Paleozoic sedimentary rock to the southeast of North Fork Palomas Creek. It cuts and offsets an older north-northwesterly trending fault about a half a mile east of Sugarloaf Peak (offset not shown on pl. 1). This older fault extends southerly to Willow Creek about 500 ft east of the northwest corner of sec. 1, T. 13 S., R. 9 W., having the same strike as the mineralized Pelican fault vein (fig. 4), discussed in the following section of this report, which appears to die out on the south side of Willow Creek, offset about 200 ft to the west, from the southern end of this older fault.

The structure and geologic units discussed in this section were obtained from preliminary field maps prepared by C. H. Maxwell and differ in some respects from those shown on the published map series I-882 covering this region (Alminas and others, 1975a, b, c).

Geochemical Information

The only available geochemical samples from this area are a few widely separated heavy-mineral concentrates from stream sediments (Alminas and others, 1975a, b, c). One such sample from a south-flowing tributary of the North Fork Palomas Creek about 1,000 ft upstream from the Dines Ranch contains more than 2,000 ppm lead; and a sample from a westerly flowing tributary where it crosses the western boundary of section 30 (about a half a mile south of Dines Ranch close to the large fault extending northeasterly from Sugarloaf Peak) contains visible fluorite. A sample from the east fork of another southerly flowing tributary near the northeast corner of section 31 contains more than 1,000 ppm tungsten. Another sample from near the mouth of a small stream draining easterly from Montosa Mountain in the northern part of section 28 close to the eastern boundary fault also contains more than 1,000 ppm tungsten, and in addition contains visible fluorite. A sample from the North Fork Palomas Creek where it crosses the eastern northerly trending range-front fault, downstream from the quartz monzonite intrusive, contains more than 1,000 ppm of both copper and tin.

Hermosa District Mineralized Area

Location

The main Hermosa mining district occupies a narrow belt on the north side of the South Fork Palomas Creek from the old town of Hermosa to the Palomas Chief mine, about 2 miles to the east (fig. 4). However, outlying mines and prospects, mineralized outcrops, and geochemical anomalies extend the area of associated mineralization northward nearly to Willow Creek along the strong northerly trending Pelican fault vein, and southward nearly to North Seco Creek in a narrow belt on the west side of the regional northerly trending range-front fault, a total distance of about 7 miles (pl. 1). These northern and southern extensions are very narrow, as is the main easterly trending district, so that the total known mineralized area is only a few square miles, in contrast to the much more extensive Chloride district area to the north.

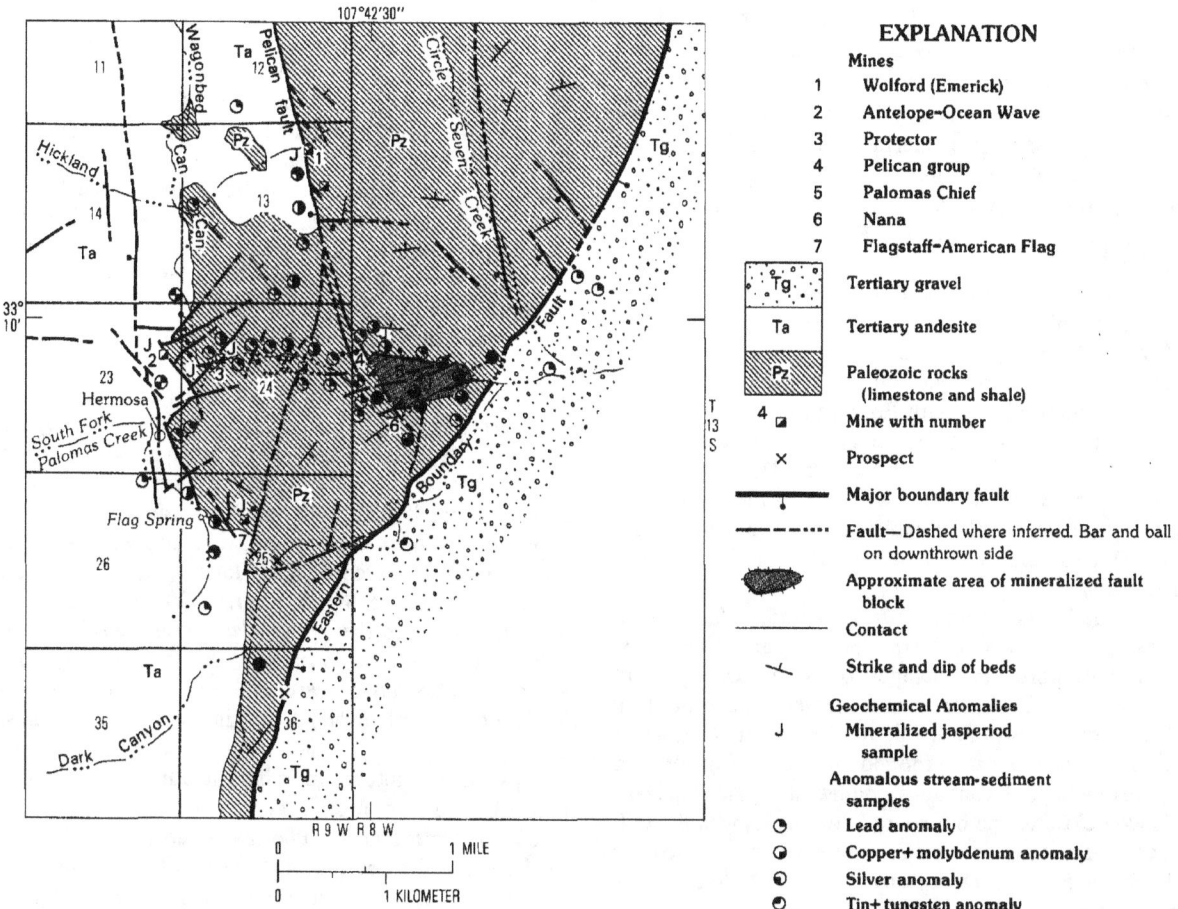

107°42'30"

EXPLANATION

Mines

1   Wolford (Emerick)
2   Antelope–Ocean Wave
3   Protector
4   Pelican group
5   Palomas Chief
6   Nana
7   Flagstaff–American Flag

Tg	Tertiary gravel
Ta	Tertiary andesite
Pz	Paleozoic rocks (limestone and shale)

4 ▫   Mine with number

×   Prospect

━━━━•   Major boundary fault

━ ━ ━•   Fault—Dashed where inferred. Bar and ball on downthrown side

▰   Approximate area of mineralized fault block

━━━━   Contact

⌐   Strike and dip of beds

Geochemical Anomalies

J   Mineralized jasperiod sample

Anomalous stream-sediment samples

◖   Lead anomaly

◑   Copper+ molybdenum anomaly

◖   Silver anomaly

◔   Tin+ tungsten anomaly

Figure 4. Sketch map of Hermosa district showing generalized location of geology, structural features, principal mines, and anomalous geochemical sample localities (modified from unpublished field map by C. H. Maxwell).

Production

Production from the Hermosa district prior to 1930 is given by Harley (1934, p. 95) as $1½ million, chiefly silver and lead with minor gold, zinc, and copper; although zinc was abundant in the ore, it was rarely recovered. There has been some recent activity in the district as a result of the increase in the price of silver, but little new ore has been produced. The largest mine was the Pelican, on the north side of South Fork Palomas Creek about a mile and a half east of Hermosa; it alone produced about a million dollars worth of ore. The Palomas Chief mine at the eastern end of the district, and the Antelope–Ocean Wave mine group at the western end, about a half a mile north of Hermosa, each produced more than $100,000 worth of ore. Other smaller mines include the Protector and Yankee Girl near the western end of the main district, the American Flag and Flagstaff mines about a mile to the south, just

east of Flag Spring, and the Wolford mine group about 1¼ miles north of the Pelican mine on the same fault vein system (Harley, 1934, p. 95–97).

Geologic Setting

Mineralization in this area appears to be confined to the Paleozoic sedimentary rocks which are exposed in a belt 2–3 miles wide west of the generally northerly trending range-front fault (fig. 4). This area extends from Willow Creek on the north to North Seco Canyon on the south, with the mineralized area mainly along the South Fork Palomas Creek east of Hermosa. Southward this belt narrows to about a mile wide east of Flag Spring. Farther south it tapers to a narrow belt less than a half a mile wide east of Dark Canyon and west of the range-front fault. This narrow belt continues southward across the divide between Dark Canyon and North Seco

Canyon. From the top of the ridge north of South Fork Palomas Creek and northward, the rocks consist largely of Magdalena Group locally overlain by patches of Bursum and Abo Formations. The canyon of South Fork Palomas Creek, east of Hermosa, has cut through the Magdalena Group and the underlying Lower Mississippian Lake Valley and Kelly Limestone; Devonian Percha Shale and Onate Formation, Silurian Fusselman Dolomite, Cutter, Aleman, and Upham Dolomite Members of the Ordovician Montoya Group. These middle and lower Paleozoic rocks below the Magdalena Group are also locally exposed in stream drainages in the narrowing belt of Paleozoic sedimentary rocks to the south of South Fork Palomas Creek.

Most of the ore in the district has come from carbonate rocks in the Fusselman Dolomite. Also yielding some ore are the lower part of the Magdalena Group beneath a black shale bed and the underlying Lake Valley Limestone above the Percha Shale (Jahns, 1955a).

The big eastern range-front boundary fault which runs southerly across North Fork Palomas Creek just west of Romero Tank changes to a southwesterly trend north of Black Peak, runs across the western end of Black Peak to cross South Fork Palomas Creek about a quarter of a mile upstream from the mouth of Dark Canyon (fig. 4). From here it continues southwesterly across Dark Canyon close to the boundary between Ranges 8 and 9 West where it swings to a more southerly course, crossing the divide between Dark Canyon and North Seco Creek near the middle of the south section line of sec. 36, T. 13 S., and R. 9 W. The major western fault which crosses the upper end of Willow Creek, west of Sugarloaf Peak, crosses Circle Seven Creek at Middle Well and runs southerly across Hickland Canyon and crosses the South Fork Palomas Creek at the town of Hermosa; it appears to die out about half a mile south of town. The major Pelican fault and vein runs roughly parallel to it, and about a mile farther east. At a point about a half a mile north of South Fork Palomas Creek in the southeast quarter of section 13, the Pelican fault splits with one branch running southerly, passing about a thousand feet east of the American Flag mine and continuing southward to Middle Seco Creek; the other branch of the Pelican fault trends southeasterly across the South Fork Palomas Creek west of the Pelican mine and dies out before it reaches the ridge crest south of the creek. Another major north-trending fault extends north from the eastern range-front fault about a thousand feet west of Circle Seven Well, roughly parallels the northerly trending portion of Circle Seven Creek, along the western flank of Brushy Mountain, crosses the southwest corner of section 32, where it swings northeasterly across the North Fork Palomas Creek, and dies out within a quarter of a mile. In the main mineralized area around the Pelican Group mines, there are several shorter northerly and northwesterly trending faults and a few easterly trending faults outlining an irregular upraised block about a half a mile long and 2,000 ft wide in the eastern part of the area, which contains most of the ore (fig. 4). In the western part, north and northeast of Hermosa, ore is largely confined to a structurally complex area between the western boundary fault and a smaller subparallel fault about 1,500 ft to the east. Smaller scattered deposits are also localized near major northerly trending faults both to the north and to the south of the main district.

Major ore bodies in the main district are structurally controlled by intersections between north-northwesterly trending faults and smaller easterly or northeasterly trending faults, or along bedding planes in the limestone near the crests of small folds. The individual ore bodies were small, producing only a few hundred tons of ore, but some were also very rich and quite numerous. According to Harley (1934, p. 94), the average oxidized ore shipped from the district ran 40% lead and 240 ounces to the ton in silver. Primary sulfide ore below the water table in the deeper eastern mine workings is also said to have contained about 40% lead and 50–100 ounces of silver per ton in some places (Jahns, 1955a).

Wall-rock alteration is intense adjacent to the ore bodies but does not extend very far from them. It consists chiefly of hydrothermal talc and coarse hydrothermal calcite derived from the host rock, with hydrothermal jasperoid locally replacing the calcite. Recrystallized hydrothermal dolomite containing disseminated pyrite commonly forms an outer alteration zone around the talc, calcite, and jasperoid.

Primary ore minerals consist principally of argentiferous galena, with subordinate chalcopyrite, bornite, and sphalerite, and minor argentite, polybasite, pyrargyrite, and tetrahedrite in a gangue of talc, quartz, calcite, clay minerals, barite, and pyrite. Close to the water table in the deeper workings is a thin zone of supergene sulfide enrichment where cavities are lined with coarse galena crystals, with secondary chalcocite, covellite, argentite, and native silver replacing primary sulfides below the water table (Jicha, 1954b). Rich oxidized ore in the upper workings consists of a vuggy, porous mass of ferruginous jasperoid associated with talc, limonite containing anglesite, cerussite, argentian plumbojarosite, cerargyrite, smithsonite, hydrozincite, hemimorphite, and commonly argentian manganese oxides, but relatively little gold or copper.

Geochemical Information

Several jasperoid samples were collected from dumps of old mines in the main district (Lovering, 1972) and closely spaced pan-concentrate samples were collected along the South Fork Palomas Creek and tributary gulches from Hermosa eastward to the mouth of Dark Canyon, with a few widely spaced samples of both jasperoid and pan concentrates representing the larger areas to the north and south of the main district (Alminas and others, 1975a, b, c; fig. 4, this report).

Rock samples.—Three jasperoid samples were collected near the western end of the district (fig. 4), one from the dump of a mine in the Antelope–Ocean Wave mining claim group about 1,000 ft north of the Hermosa cemetery, one from an outcrop at the junction of northeasterly and northwesterly trending faults on a ridge west of the mouth of Hickland Canyon about 1,000 ft east of the cemetery, and one from an outcrop along the northeast-trending fault close to the Protector mine about 1,500 ft to the northeast. Another sample of barren jasperoid was collected from a major northerly trending fault zone in limestone on the south side of South Fork Palomas Creek about a half a mile east of the mouth of Hickland Canyon about midway between the eastern and western parts of the district. Four samples were taken from the eastern (Palomas Camp) end of the district, consisting of dump samples from the upper east shaft of the Pelican claim group, from near the center of the Pelican group of claims, and also from the dump of the Palomas Chief mine at the east end; one outcrop sample of unmineralized jasperoid was collected from a northeast-trending fault zone with Lake Valley Limestone on the southeast and Magdalena Group on the northwest, on a ridge about 300 ft east of the Palomas Chief mine.

All the dump samples consist of dense aphanitic, medium-gray to dark-gray jasperoid. Those from dumps at the east end of the district contain primary sulfide minerals including galena, chalcopyrite, sphalerite, and pyrite, with coarse late calcite gangue. The dump sample from the Antelope-Ocean Wave claims at the west end of the district contains hematite, limonite, and calcite. The jasperoid outcrop samples are mostly shades of brown, reddish brown, and grayish red with an aphanitic texture. Samples from the outcrop on the major north-trending fault zone on the south side of South Fork Palomas Creek also have an aphanitic texture, but display a variety of colors including grayish red, reddish purple, dusky yellow green, olive gray, yellowish gray, dark greenish gray, light gray, and grayish orange. Most of the dump samples are anomalously high in silver (60–900 ppm), those from dumps at the eastern end of the district are also anomalous in lead (1,000–1,500 ppm), copper (700–1,500 ppm), zinc (700–1,500 ppm), molybdenum (15–30 ppm), and, locally, cadmium (50 ppm). Outcrop samples of jasperoid from the Hermosa district are more variable in composition, but all outcrop samples are consistently richer in vanadium and lower in silver than

the dump samples. The sample from the fault junction near the Flagstaff mine in the southwestern part of the district contains anomalous tungsten (150–1,000 ppm) and arsenic (±700 ppm) and contains up to 70 ppm germanium, 7 ppm beryllium, 100 ppm niobium, and 1,500 ppm antimony. The sample from the outcrop near the Protector mine is rich in manganese (0.5%) and cadmium (50 ppm), and slightly anomalous in barium (50 ppm), chromium (30 ppm), gallium (5–10 ppm), germanium (10 ppm), and zinc (500 ppm). The samples from the north-trending fault zone on the south side of South Fork Palomas Creek are anomalously high in cobalt (10–15 ppm) and nickel (70 ppm); those from the northeasterly trending fault zone at the eastern end of the district are also anomalous in cobalt (7 ppm), chromium (300 ppm), nickel (20–30 ppm), and yttrium (15–20 ppm).

Outlying jasperoid samples, outside the main district, have been collected from the following localities: the dump of the Wolford mine, about a mile and a half to the north on the Pelican fault vein; the dump of the American Flag mine, about a mile south of South Fork Palomas Creek, east of Flag Spring; an outcrop on a north-northeast-trending range-front fault in a saddle at the western end of Black Peak, about a mile and half northeast of the Palomas Chief mine, on the south boundary of the Sugarloaf Peak 7½′ quadrangle; an outcrop on a northerly trending fault zone just east of the road in the pass between Dark Canyon and North Seco Canyon, about 2½ miles south of Hermosa; and on the north side of North Seco Canyon opposite Davis Well, about a mile farther south of the last locality at the southern edge of the Hermosa mineralized area. Samples from the Wolford mine dump representing ore zone jasperoid replacing Magdalena Group near the Pelican fault vein are dense, fine grained to aphanitic with disseminated galena and calcite gangue; they are anomalous in silver (30–400 ppm), copper (300 ppm), lead (200–2,000 ppm), and zinc (0.1–0.7%), slightly anomalous in molybdenum (15–30 ppm), and one sample contains 0.1 ppm gold. Samples from the dump of the American Flag mine represent a silicified breccia zone in Lake Valley Limestone along a north-trending fault. They consist of angular fragments of aphanitic, white to yellowish-gray chert in a matrix of fine- to medium-grained medium-gray jasperoid containing relict sulfides (galena, sphalerite, chalcopyrite, and pyrite) as well as anglesite, cerussite, hemimorphite, argentian plumbojarosite, and abundant orange goethite. Samples contain as much as 700 ppm silver, 300 ppm cadmium, 1,500 ppm copper, 1,500 ppm lead, 1.5% zinc, 700 ppm manganese, 30 ppm molybdenum, 20 ppm vanadium, and 15 ppm yttrium. Outcrop samples from the fault zone west of Black Peak, with silicified Fusselman Dolomite limestone to the west, consist of aphanitic grayish-red to medium-gray breccia fragments cemented by a matrix of fine- to medium-grained, locally vuggy, medium-dark-gray jasperoid, and the samples are slightly anomalous in silver (3 ppm), beryllium (2 ppm), and tungsten (50 ppm). A sample from a jasperoid reef along a northerly trending fault zone in lower Paleozoic (possibly Lake Valley Limestone) carbonate rock east of the road on the south boundary of the Sugarloaf Peak 7½' quadrangle, consisting of medium-dark-gray aphanitic breccia fragments in a fine-grained light-greenish-gray matrix, contains talc, sericite, fluorspar, and pyrite, plus manganese oxide, hematite, jarosite, and goethite veinlets. It contains as much as 5% magnesium, 40 ppm silver, 300 ppm lead, 50 ppm vanadium, 9% fluorine, 2,000 ppm zinc, 700 ppm manganese, and 70 ppm copper. The southernmost outcrop sample from North Seco Creek consists of mottled aphanitic grayish-brown and moderate-brown jasperoid in limestone, close to its contact with Tertiary volcanics along a north-trending fault zone. It contains up to 300 ppm vanadium, 30 ppm silver, 1,000 ppm lead, 5 ppm cobalt, 70 ppm molybdenum, and 500 ppm zinc.

Stream-sediment concentrates. —Numerous anomalous heavy-mineral concentrates from stream sediments have been collected east of Hermosa townsite along and near the South Fork Palomas Creek to its junction with Dark Canyon, also along and near lower Hickland Canyon, and the north-flowing tributary north of Flag Spring (Alminas and others, 1975a, b, c). A continuous belt of high lead anomalies (greater than 2,000 ppm) locally with silver anomalies extends from Flag Spring northward to South Fork Palomas Creek about parallel to the trend of the major fault to the east and also eastward along South Fork Palomas Creek across the boundary fault to the mouth of Dark Canyon, also for about a half a mile upstream from the mouth of Hickland Canyon and the south-flowing tributary of Palomas Creek about ¾ mile east of Hickland Canyon. Silver anomalies (greater than 100 ppm) are concentrated in the eastern part of the area from below the westernmost Pelican mine workings downstream about ¾ mile to a point about 500 ft upstream from the eastern boundary fault; these include several compound anomalies (Cu > 1,000 ppm; Pb > 2,000 ppm; Zn > 10,000 ppm; Ag > 100 ppm; and Mo > 70 ppm). Molybdenum concentrations greater than 70 ppm were found in samples from a tributary that drains southward into South Fork Palomas Creek ¾ mile east of Hickland Canyon and also from the north-draining tributary of South Fork Palomas Creek west of Flag Spring. Three isolated anomalies for lead (> 2,000 ppm) occur in T. 13 S., R. 9 W., one just east of the junction of Wagonbed and Hickland Canyons, another in SE¼ sec. 12 on southwesterly draining tributaries of Wagonbed Canyon, and another in SE¼ sec. 1 on an easterly flowing

tributary of Circle Seven Creek, all apparently related to mineralization along the long northerly trending Pelican fault vein. Isolated lead anomalies were also found near lower Circle Seven Creek south of Brushy Mountain. A compound anomaly (zinc > 10,000 ppm; lead > 2,000 ppm; silver > 100 ppm; and molybdenum > 70 ppm) was found in concentrates from a short west-draining gulch near the middle of the north boundary of sec. 36 (T. 13 S., R. 9 W.), on a north-trending fault zone just south of Dark Canyon about a mile south of the American Flag mine.

Summary and evaluation of geochemical samples. — Most of the mine dump samples of jasperoid contain anomalous concentrations of silver and base metals, particularly lead. However, few of the outcrop jasperoid samples do, although they are consistently high in vanadium and nickel. This suggests possible leaching of base and precious metals from jasperoid outcrops in the mineralized area. In the pan-concentrate samples, the best indicator element is lead, which gives strong anomalies downstream for at least a half a mile from the mineralized source; molybdenum anomalies also yield a fairly extensive downstream anomaly train. Silver anomalies and compound anomalies for ore elements other than lead and molybdenum are found only in pan-concentrate samples within a few hundred yards of their source mineralization. Major northerly trending fault zones show evidence of mineralization for 2 or 3 miles, both to the north and to the south of the main Hermosa district, indicating a strong possibility of outlying buried ore deposits related to these structures.

North Seco Canyon–Cave Creek Gap

Another apparent gap or break in the western mineral belt extends southward from North Seco Canyon across Las Animas Creek to Cave Creek, a distance of about 7 miles. The geology of this remote and poorly accessible area has not been studied in detail. The following information is taken from an unpublished preliminary field map prepared by C. H. Maxwell, and differs in some respects from the published map I-882 (Alminas and others, 1975a, b, c).

The big eastern range-front boundary fault, which separates younger gravels to the east from older rocks to the west, crosses North Seco Canyon about half a mile east of Davis Well, and continues southward about 2 miles to to the divide between Turkey Canyon and Conley Canyon northeast of Conley Spring. A narrowing belt of lower Paleozoic rocks on the western side of this fault is about ½ mile wide where it crosses North Seco Canyon, and pinches out against the boundary fault on the north side of Middle Seco Canyon, with Tertiary volcanic rocks to the west. The northerly trending segment of the range-front boundary fault terminates against a strong northwesterly trending fault, which marks the boundary between younger gravels to the east and Tertiary volcanic rocks to the west. This northwesterly trending fault continues southeastward across the junction of Conley Canyon and South Seco Canyon to the north side of Hyler Canyon just east of the western boundary of sec. 19, T. 14 S., R. 8 W., where a northeasterly trending fault from the south is cut off by it and it continues to Las Animas Creek. The northeasterly trending fault changes to a northerly trend along the upper eastern fork of Hells Canyon and forms the boundary between younger gravel to the east and bedrock to the west, to the point where it crosses Las Animas Creek approximately half a mile downstream from the Dunn Place; it continues southward. Another northwesterly trending fault along Las Animas Creek shifts the gravel-bedrock boundary southeast to the confluence of Las Animas Creek with a small northerly draining tributary about 0.2 miles southwest of Apache Camp, where a strong north-trending fault comes in and forms the western boundary fault from this junction southward on the eastern side of Apache Peak, crossing Cave Creek about ¼ mile downstream from its confluence with Cave Canyon. These faults are shown on plate 1. Another strong northerly trending fault, downthrown to the west, crosses Las Animas Creek about half a mile upstream from the Dunn Place and runs southerly on the east side of Magner Park, crossing upper Cave Creek about half a mile north of Magner Spring. On the upraised block to the east of this fault, the Tertiary volcanics have locally been stripped off, exposing Paleozoic sedimentary rocks beneath, in a belt about a quarter of a mile wide south of the Dunn Place on Las Animas Creek, southward to the fault west of Apache Peak.

The North Seco Creek–Cave Creek area may have valuable mineral deposits in it. C. H. Maxwell (oral commun. to Heyl, 1980) has found rich silver mineral float in the area, but no geochemical rock samples or highly anomalous stream-sediment-concentrate samples have been collected in this interval between North Seco Canyon and Cave Creek, and there are no mines or prospects in this area. West of the boundary fault zones, relatively few heavy-mineral concentrates have been collected from this area. One such sample from the mouth of a north-flowing tributary of North Seco Canyon east of Davis Well contains >1,000 ppm zinc and >2 ppm silver Another from the upper east fork of Hells Canyon, near the boundary fault, contains >1,000 ppm zinc. Zinc anomalies in upper Turkey Canyon and copper anomalies on north-flowing tributaries of Las Animas Creek, were also found east of the boundary faults.

Kingston District, Including North Percha Creek and South Percha Creek Areas

Location

The western flank of the Black Range is almost continuously mineralized in a belt about 10 miles long and from 1 to 3 miles wide extending from Cave Creek southward to South Percha Creek. This area is part of the Kingston district and includes the North Percha Creek area at the north end, the main Kingston district in the middle, and the South Percha Creek area at the south end, covering about 20 square miles (fig. 5). North of North Percha Creek to Cave Creek, the mines and prospects diminish and become less important, but include such mines as Barite Hill and Ingersol near the northwest head of Dumm Canyon. The more important mines of the North Percha Creek area lie south of Dumm Canyon along North Percha and Mineral Creeks, including the very well-known Virginia and many adjacent, productive mines.

Production

The Kingston district has produced close to $7 million worth of ore, mostly silver and lead, with subordinate copper, zinc, manganese, tungsten, and gold (Harley, 1934, p. 102–103). Most of the activity in the district took place in the late 1800's prior to 1895. All the major mines in the northern and central parts of the area were abandoned and inaccessible when the first geologic report on the area was written (Lindgren, Graton, and Gordon, 1910); consequently, little is known of the mineralogy of the rich supergene ores. The main producing mines in the North Percha Creek area, including the famous Virginia mine, produced mostly silver with some gold, base metals, and tungsten. Mines in the main Kingston district, south of Mineral Creek and Middle Percha Creek, included the Blackeyed Susan, Andy Johnson, Brush Heap, Calamity Jane, United States, and Illinois in the western part, and the Comstock, Black Colt, Kangaroo, Caledonia, Lady Franklin, Superior, Bullion, Gypsy, Iron King, and General Jackson in the northern and eastern parts. Mines in the western part produced abundant silver, lead, and zinc with some copper; those to the north and east produced mainly silver (Harley, 1934, p. 103–105). There is only one major mine group in the South Percha Creek area, the Gray Eagle mine group which produced silver, lead, copper, gold, and zinc (Hill, 1946).

Geologic Setting

The geology of this mineralized area is quite complex (fig. 5). Most of the mines in the northern part

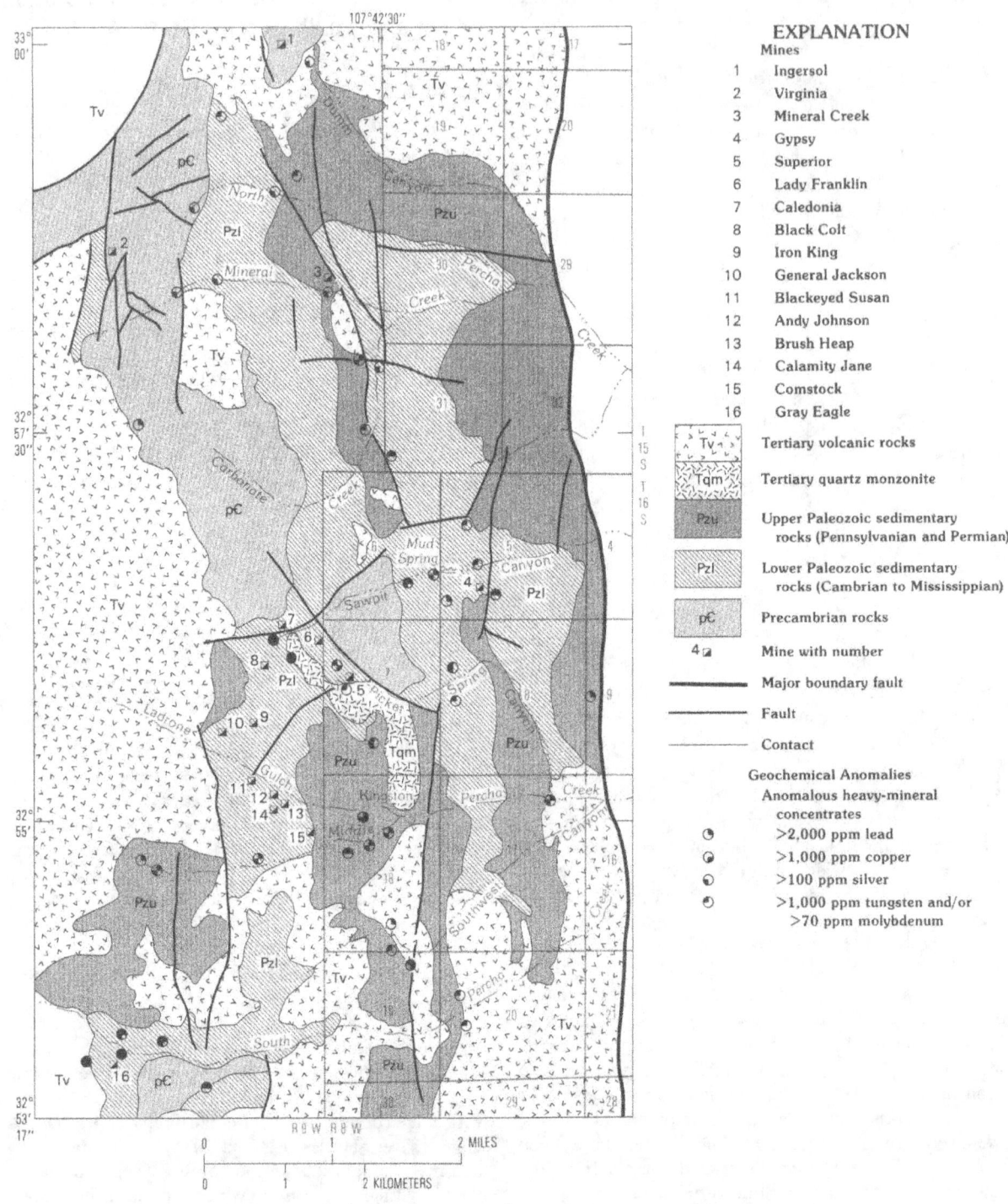

Figure 5. Sketch map of Kingston district, showing general geology, approximate location of structural features (modified from Hedlund, 1977a), principal mines, and anomalous geochemical sample localities.

of the area are along strong northerly trending fault veins cutting Precambrian granitic rocks; a few are along westerly trending cross faults; and some of the smaller mines along lower Carbonate Creek in the eastern part of the North Percha Creek area are in north-trending veins cutting Lake Valley Limestone. To the south of Sawpit Canyon along upper Picket Spring Canyon and Ladrone Gulch, southward to the valley of Middle Percha Creek in the western part of the Kingston district, is a strongly mineralized belt in lower Paleozoic rocks south of the westerly trending Sawpit fault, and southwest of the northwesterly trending Picket Spring fault on the western side of an altered monzonitic intrusive body (fig. 5). Ore bodies are largely in Silurian Fusselman Dolomite and Ordovician Montoya Group in carbonate rocks below the impermeable Percha Shale (Devonian) close to the intrusive contact. To the east of the monzonitic intrusive, the section is repeated along a strong northerly trending fault zone, and most of the mines here are also in Ordovician carbonate rocks beneath Percha Shale close to the faults. Some ore in the western part of the district has been produced from the El Paso Limestone below the Montoya Group, and also from the overlying Lake Valley Limestone and Magdalena Group in the eastern part (Harley, 1934, p. 100–101). To the west of the mineralized area is a thick pile of barren Tertiary volcanic rocks overlying Precambrian granite and the Paleozoic sedimentary rocks. In the southern part of the area, between Middle and South Percha Creeks west of Kingston, large masses of younger Tertiary rhyolitic rocks cover much of the underlying Paleozoic sedimentary section. At the only important mine in this area, the Gray Eagle, ore occurs in the El Paso Limestone in close proximity to intrusive andesite and rhyolite dikes (Hill, 1946). Although similar dikes are also present to the north in the main Kingston district, they do not appear to be related to the ore deposits in the main area.

Major faults continue to trend northerly in this area. The eastern range-front boundary fault appears to run southerly across the junction of Dumm Canyon and North Percha Creek and crosses Middle Percha Creek about a half a mile east of Kingston Ranger Station (pl. 1). North of Dumm Canyon, early Tertiary volcanic rocks are exposed on the west side of this fault; from this point south to Middle Percha Creek, the Magdalena Group forms the western block (fig. 5). Tertiary volcanic rocks and gravel interspersed with local exposures of Mississippian and Pennsylvanian carbonates (Lake Valley and Magdalena Group) are present on the western side of the boundary fault. Another major northerly trending fault parallels the range-front boundary fault about a mile and a half to the west along upper Dumm Canyon, across a saddle between North Percha and Mineral Creeks about a mile west of their

junction, down lower Carbonate Creek where it appears to terminate north of Sawpit Canyon, but the fault seems to reappear just south of Picket Spring Canyon, crossing Middle Percha Creek at Kingston and, farther south, crossing South Percha Creek about 2½ miles east of the Gray Eagle mine (fig. 5). Precambrian rocks are exposed on the western side of this fault west of Dumm Canyon, and also in the area between Sawpit and Picket Spring Canyons where this fault seems to disappear; elsewhere, north of Kingston, Paleozoic sedimentary rock is on both sides of it; to the south of Kingston, the fault cuts Tertiary volcanic rocks. Several mines are located along this fault from Kingston northward to McCann Gulch. Several other strong northerly trending faults, upthrown to the west, are also present to the west of this one, between Carbonate Creek and Dumm Canyon; they are also mineralized locally in this part of the area. Another north-trending fault with Tertiary volcanic rocks to the west marks the western boundary of the southern part of the Kingston district; the fault extends from Ladrone Gulch, across Middle Percha Creek about 2 miles west of Kingston, and dies out on South Percha Creek about ¾ mile east of the Gray Eagle mine. South of Middle Percha Creek, the Magdalena Group and the Abo Formation are extensively exposed on the west side of this fault, but the lower Paleozoic rocks on the east side are covered by younger Tertiary volcanic flows (Hedlund, 1977a). Several shorter easterly or northeasterly trending, locally mineralized, cross faults also are present north of Middle Percha Creek, and a strong northwesterly trending fault occupies upper Picket Spring Canyon, crossing a saddle and dying out north-westward in Precambrian rocks on the ridge between Sawpit Canyon and Carbonate Creek. These cross faults are largely limited to the central part of the mineralized area, between Middle Percha and North Percha Creeks. Ore bodies are localized by breccia zones at fault intersections and changes of strike or dip along fault zones. In the main Kingston district, ore bodies also occupy the crests of small anticlinal folds in carbonate rocks beneath the Percha Shale (Lindgren, Graton, and Gordon, 1910; Harley, 1934, p. 100–102).

Wall-rock alteration consists of a narrow zone of propylitic and sericitic alteration adjacent to the veins in the Precambrian granitic rocks. In the Paleozoic sedimentary rocks, both the carbonates and overlying shale beds were extensively silicified along faults and fractures by early pre-ore solutions. The ore stage of mineralization was marked by the introduction of talc, clay minerals, and sericite surrounding the ore bodies. The monzonitic intrusive in the Kingston district produced very little contact-metasomatic alteration in the sedimentary rocks, although the intrusive is itself strongly hydrothermally propylitized, argillized, and sericitized. Some of this alteration may be caused by the large caldera west of the district.

The quartz veins in Precambrian rocks in the northern part of the area contained wolframite and ferberite in the tungsten deposits. Galena, sphalerite, and chalcopyrite with minor polybasite and chalcocite were the primary ore minerals in the base-metal-and-silver vein deposits. Cerargyrite was abundant in the oxidized zone, where it was locally associated with native silver. The main gangue minerals were quartz, barite, and pyrite or limonite. The breccia zone replacement and pod-form ore bodies in carbonate rocks contained argentiferous galena, sphalerite, chalcopyrite, and argentite as primary ore minerals in a gangue of rhodochrosite, rhodonite, calcite, quartz, and pyrite. These minerals are rare and remain only in the deepest workings. In the oxidized zone where most of the ore was produced, supergene ore minerals included cerargyrite, argentojarosite, native silver, anglesite, cerussite, smithsonite, malachite, azurite, and cuprite in a gangue of limonitic clay and argentian manganese oxides, all encased in talc envelopes in the carbonate rocks. In some places the manganese oxides were sufficiently abundant to be mined as manganese ore.

Geochemical Information

Although no jasperoid samples are available from this area, if Harley's observation that they represent a barren pre-ore alteration is correct (Harley, 1934, p. 101), such samples would be unlikely to provide ore-related geochemical anomalies. Heavy-mineral concentrates from stream-sediment samples show numerous strong metal anomalies at many localities in this mineralized area all the way from McCann Gulch on the north to South Percha Creek (Watts, Alminas, and Kraxberger, 1978; Watts and others, 1978a, b; Alminas and others, 1977a, b, c, d; fig. 5, this report). Although many of these may be contamination from old mine dumps, they indicate the metals derived from known mineral deposits. A strong bismuth-lead anomaly was found on McCann Gulch just north of Bald Hill, and a silver anomaly was found near the mouth of a north-flowing tributary of McCann Gulch about 1,000 ft southwest of the junction of McCann Gulch and Cave Creek. Several isolated silver anomalies were also found along North Percha Creek and Mineral Creek, and a few tungsten and bismuth anomalies on small tributaries north of North Percha Creek and west of Dumm Canyon. Most of these anomalies are probably derived from the northerly trending mineralized veins in this area. In the mineralized fault zone along lower Carbonate Creek, there are lead, silver, and molybdenum anomalies, and one tin anomaly. Farther south, in Sawpit Canyon, compound copper-lead-zinc-silver-molybdenum anoma-

lies were found in the upper part of the canyon, which may represent contamination from old mines in the west Kingston district. Molybdenum anomalies continue downstream to the vicinity of Mud Spring, and some are also associated with lead and silver anomalies. Molybdenum anomalies, locally associated with lead, silver, and tungsten anomalies, also are present along Picket Spring Canyon and its tributary washes all the way to its confluence with Middle Percha Creek. These anomalies may be related to the small monzonitic stock south of Picket Spring Canyon. Lead and silver anomalies also occur along Middle Percha Creek at its confluence with Ladrone Gulch and for about 2 miles upstream. A compound copper-lead-zinc-silver anomaly in the lower part of Ladrone Gulch probably reflects the known mineralized area in the upstream drainage. Copper-lead-tungsten-molybdenum anomalies are present along a north-trending fault zone on Southwest Canyon and its tributary washes about a mile south of Kingston, which may reflect the presence of new deposits since there are no mines in this area. On South Percha Creek there are strong compound copper-lead-zinc-silver-tungsten and molybdenum anomalies close to the Gray Eagle mine workings. Lead-silver-tungsten anomalies persist for about a half a mile downstream, and molybdenum anomalies for about 3 miles downstream.

Summary and evaluation of geochemical samples.—Lead and silver anomalies in stream-sediment heavy-mineral concentrates are widely distributed throughout this mineralized area. At the northern end north of Mineral Creek, there are also several tungsten and bismuth anomalies. From lower Carbonate Creek southward to South Percha Creek, there are numerous compound tungsten-molybdenum anomalies, although neither of these metals was recovered from the old mines in the Kingston district. Streams draining the western part of the area south of Carbonate Creek also yielded scattered copper and zinc anomalies. The widespread tungsten-molybdenum anomalies could reflect the presence of undiscovered scheelite and molybdenite deposits.

South Percha Creek–Trujillo Canyon Gap

Another short break in the western mineral belt appears to exist between South Percha Creek and Trujillo Canyon, about 1½ miles to the south across a high easterly trending spur of the Mimbres Mountains. This gap contains extensive exposures of lower Paleozoic sedimentary rocks with an isolated mass of Precambrian granite projecting northward into the southwestern part of the area, about a mile south of the Gray Eagle mine, the southernmost mine of the Kingston district. Paleozoic sedimentary rocks are locally capped by younger Tertiary rhyolitic volcanic rocks along the ridge crest. These older

Paleozoic rocks, consisting largely of Percha Shale and underlying Ordovician carbonate rocks, are cut off to the east by a northerly trending fault which extends southward from South Percha Creek about a mile downstream from the Gray Eagle mine, which produced silver, lead, gold, and copper. To the east of this fault, isolated masses of Magdalena Group locally show through the overlying younger Tertiary volcanic rocks. Another small isolated block of older Precambrian mafic rocks is exposed west of this northerly trending fault on the south side of South Percha Creek to within about a quarter of a mile of the Gray Eagle mine. One anomalous heavy-mineral concentrate containing high concentrations of lead, tungsten, and molybdenum was taken from a short northerly trending tributary of South Percha Creek, about 1,000 ft south of the creek, ¾ mile east of the Gray Eagle mine, and close to the southern boundary of these Precambrian rocks (Alminas and others, 1977c; Watts and others, 1978a, b).

Tierra Blanca District—Western Part

Location

The western part of the Tierra Blanca district is about 5 miles south-southwest of Kingston in the upper drainage basins of Trujillo Canyon and Tierra Blanca Creek and on Seven Brothers Mountain to the west. It covers an area about 2 miles long from north to south and a mile wide from east to west.

Production

This district produced gold, silver, and lead ore with a total value of about $150,000 in the late 1800's. Most of this came from the Lookout mine on upper Trujillo Canyon at the northern end of the district (Harley, 1934, p. 108). Smaller mines to the south and southwest include the Big Jap and Beardog.

Geologic Setting

The western part of the district area is complex geologically, with a number of small blocks of lower Paleozoic rocks and the underlying Precambrian granite locally exposed along small closely spaced northerly trending faults (Hedlund, 1977a). A major north-trending fault runs along the eastern base of Seven Brothers Mountain with Precambrian granitic rock to the west. This fault swings southwesterly at Tierra Blanca Creek along the upper valley of that stream. Many hills in the area are capped with late Tertiary white rhyolite

flows, from which the district gets its Spanish name. Near the heads of Trujillo Canyon and Tierra Blanca Creek, east of the mineralized area, Lake Valley Limestone and the Magdalena Group are exposed in a mile-wide belt west of a north-trending fault with younger Tertiary rhyolite on the east side. Ore is largely confined to silicified Bliss Sandstone and the overlying El Paso Limestone, though locally ore extends upward into Fusselman Dolomite. It is localized by proximity either to rhyolite dikes and sills or to small northerly or northeasterly trending faults. Ore bodies are in gash veins or tension fractures in the silicified sandstone or limestone, and bedding plane replacements in the limestone. High-grade gold-telluride ore is confined to the Bliss Sandstone, with lower grade silver-lead sulfide ore in the overlying carbonate rocks. Free gold and hessite occur in a gangue of amethystine quartz in the quartzitic sandstone. Galena, sphalerite, chalcopyrite, with subordinate argentite and acanthite in a gangue of pyrite, quartz, and calcite characterize the ore bodies in the overlying carbonate rocks (Harley, 1934, p. 107). The Fusselman Dolomite beneath the Percha Shale has been extensively silicified in this area, forming a broad but thin blanket of jasperoid, which has been brecciated and recemented by pink or white vein quartz locally. Although this jasperoid locally contains traces of gold in the breccia cement, it is subeconomic in grade (Lovering, 1972).

Geochemical Information

Heavy-mineral concentrates from stream-sediment samples collected in Trujillo Canyon about a quarter of a mile below the Lookout mine in the western part of the Tierra Blanca district gave strong anomalies for silver, lead, tungsten, and molybdenum. Another sample, about a half a mile below the mine showed the same anomalous metals except for silver. A sample about a mile downstream from the mine showed anomalies only for lead and molybdenum, and a final sample, 1½ miles below the mine, was anomalous in molybdenum only. Other heavy-mineral concentrates from stream sediments below the smaller mines on upper Tierra Blanca Creek in the southern part of the western area also showed compound lead-tungsten-molybdenum anomalies, but no silver (Alminas and others, 1977c; Watts and others, 1978a, b). Unfortunately, no jasperoid samples are available from this area.

Stoner Mountain–Juniper Peak Gap

Beyond the western part of the Tierra Blanca district, the western mineral belt appears to be offset about 3½ miles to the east, from the western flank of Stoner Mountain to the gulch that drains southward into Tierra Blanca Creek, east of Juniper Peak and northwest of Tierra Blanca Mountain. Most of this interval is covered by older Tertiary mafic volcanic rocks, which are overlain by younger Tertiary rhyolitic volcanics between the eastern flank of Stoner Mountain and the western flank of Juniper Peak. In the western part of this gap, these younger volcanics have been eroded in places, revealing Pennsylvanian limestone of the Magdalena Group beneath; they are cut off toward the east by a northerly trending fault which crosses Tierra Blanca Creek about a mile east of Stoner Mountain. There are no mines or prospects, and no heavy-mineral-concentrate anomalies in this gap. However, to the south and southeast of the Tierra Blanca district—western part, west and southwest of Pine Spring Mountain on upper Berenda Creek[2] and its southerly flowing tributaries, panned heavy-mineral-concentrate samples containing visible fluorite are widely distributed; and samples from the upper basin of Macho Canyon to the south contain scattered tin anomalies as well as visible fluorspar (Watts, Alminas, and Kraxberger, 1978).

Tierra Blanca District—Eastern Part

Location

The eastern part of the Tierra Blanca district extends from the vicinity of the Log Cabin mine about a mile north-northwest of Tierra Blanca Mountain, around the northern and eastern flanks of this mountain, across the divide between lower Stoner Creek and Silvertail Canyon, then southerly down Silvertail and Pierce Canyons to about the mouth of Cottonwood Creek. It is about 5 miles long north-south with a maximum width of about a mile in the northern part.

Production

This area has a total recorded production of about $150,000, prior to 1900. Most of this was silver and lead, although some gold and copper were also produced. The major mines include the Log Cabin at the north end, the Hornet group on Tierra Blanca Creek northeast of

---

[2]Although Berenda Creek and Berenda Mountain are spelled "Berrenda" (two r's) on several U.S. Geological Survey topographic maps, including the 1935, 15-minute Hillsboro map, we will use the actual family name in this report—Berenda with one "r." This follows the local usage of the area as determined by C. H. Maxwell (oral commun., July 1988).

Tierra Blanca Mountain, the Jay Hawk mine group south of the mountain on Stoner Creek, and the Silvertail mine group on Silvertail and upper Pierce Canyons (Harley, 1934, p. 108).

Geologic Setting

All the mines in this part of the Tierra Blanca district are concentrated in the lower Paleozoic carbonate rocks of Ordovician and Silurian age beneath the Devonian Percha Shale. These form a narrow northerly trending belt faulted down against Tertiary volcanic rocks to the west, with Lake Valley Limestone and Magdalena Group in normal sequence to the east from Cottonwood Creek northward to Tierra Blanca Creek. Here, at Tierra Blanca Creek, the belt is cut off and may be shifted westerly about a mile along Tierra Blanca Creek to a point about a mile east of Juniper Peak where it resumes a northerly trend for about a mile to Trujillo Canyon. Although the belt extends on northward for some distance, the mineralization appears to be largely limited to that part of it between Trujillo Canyon and Tierra Blanca Creek. This northern part of the district is bounded on the east by the southern end of the same large range-front boundary fault, downthrown to the east, which crosses Middle Percha Creek at Kingston Ranger Station, the fault dies out to the south in a complex brecciated zone characterized by numerous small northerly, northeasterly, and northwesterly trending faults on the north side of Tierra Blanca Creek. A short, more or less easterly trending, fault system at the northern end of Tierra Blanca Peak shifts the mineralized belt eastward about a mile. From this point south, the mineralized belt follows another strong northerly trending fault with Tertiary volcanic rocks faulted downward on the western side against the lower Paleozoic sedimentary rocks to the east. The ore is in veins and breccia zones along northerly trending faults, commonly in the vicinity of rhyolite and andesite dikes (Harley, 1934, p. 107). It also forms small replacement mantos beneath Percha Shale at the Silvertail mine. Wall-rock alteration consists of silicification along the mineralized faults, pyritic alteration in the dikes, and talc envelopes around the ore bodies. The ore consists of galena, chalcopyrite, argentite, and acanthite with supergene cerussite and cerargyrite in a pyritic quartz gangue.

Geochemical Information

A strong compound geochemical anomaly for copper, lead, silver, molybdenum, and tungsten is present in heavy-mineral concentrates near the mouth of the south-draining tributary of Tierra Blanca Creek below the Log Cabin mine. Other anomalies for silver and for lead and molybdenum were found near the mouths of tributary gulches east and northeast of Juniper Peak. Lead and molybdenum anomalies are present for about 2 miles down Tierra Blanca Creek to just above the J.P. Nunn Ranch. Heavy-mineral concentrates from Stoner Creek on the south side of Tierra Blanca Mountain show silver and molybdenum anomalies for about a half a mile downstream below the Jay Hawk mine group. Samples from Silvertail Canyon, below the Silvertail mine, show compound anomalies for silver, lead, copper, molybdenum, and tungsten. Lead and molybdenum anomalies were also found downstream in Pierce Canyon above the mouth of Cottonwood Creek. Just below this stream junction a heavy-mineral concentrate at the mouth of a short westerly draining tributary of Pierce Canyon shows a compound lead-silver tungsten-molybdenum anomaly. Scattered anomalies for lead, molybdenum, and tungsten were also found in samples taken farther south, to the north, east, and southeast of Signal Peak (Alminas and others, 1977a, b, c; Watts, Alminas, Kraxberger, 1978; Watts and others, 1978a, b) following the northwest extension of the Lake Valley fault. However, the eastern part of the Tierra Blanca district mineralized area appears to end a short distance east of the east side of Signal Peak.

Although there are a few widely scattered isolated copper anomalies in heavy-mineral concentrates from localities farther south, there are no mines farther south, except a small prospect near Grapevine Spring about 9 miles to the south. The stream junction of Pierce Canyon and Cottonwood Creek appears to mark the southern terminus of the western mineral belt, about 12 miles north of the Grant County line.

GENERAL SUMMARY OF THE EASTERN MINERAL BELT

The eastern mineral belt extends from the northern end of the Sierra Cuchillo on Alamosa Creek about 6 miles north of the Sierra-Socorro County line (north of pl. 1 top edge), southward through the Iron Mountain mineralized area on the northwest flank of the Sierra Cuchillo in southern Socorro and northern Sierra Counties, to about 50 miles southward to the Macho district just north of Greg Hills in southern Sierra County about 2 miles north of the Luna County line. In Sierra County the belt includes, from north to south, the Iron Mountain mineralized area (now called district; Jahns, 1944), the Hanson fluorite vein north of Red Hill in the northern part of the Cuchillo Negro district, the Chise fluorite mineralized area, the southern Salado Mountains mineralized area, the Hillsboro district including Copper Flats and the Golddust Camp placers all in the Animas Hills (Harley, 1934, p. 19, p. 168), the Lake Valley district, and the Macho district (pl. 1). This belt, like the western belt, is also discontinuous with many wide gaps of unmineralized or sparsely mineralized intervals or unprospected areas, particularly in the central and southern parts of the belt. Its western boundary is approximately 6 miles east of the western belt in the northern and southern portions and about 15 miles east of the western belt in the central portion.

The eastern belt differs topographically from the western belt. The northern third lies within the Sierra Cuchillo, a well-defined basin-and-range-type mountain range, which terminates a few miles south of Chise. Southward from this point the belt crosses a broken upland drained by easterly flowing tributaries of the Rio Grande and interrupted at irregular intervals by isolated groups of hills, such as the Garcia Peaks, Salado Mountains, Animas Hills, Sibley Mountain, Apache Hill, and Town Mountain.

The total value of mineral production from the eastern mineral belt is not known accurately but is probably in the neighborhood of $15 million, most of which is accountable to gold from the mines of the Hillsboro district and to small, but very rich, silver mines of the Grande Group in the southern part of the Lake Valley district. The only individual large mine with a recorded production of more than a million dollars is the Rattlesnake mine in the Hillsboro district (Harley, 1934, p. 141).

A little iron ore has been produced from a ferruginous tactite zone at Iron Mountain near the northern end of the belt in Sierra County and small quantities of tungsten, fluorspar, zinc, and beryllium associated with the iron ore have been mined. There are also some small copper, uranium, and lead mines and fluorspar prospects in the Paleozoic carbonate rocks south and southeast of the iron deposits. Copper ore with subordinate lead, zinc, fluorite, gold, and silver was produced from contact-metasomatic tactite deposits in several small mines in the Cuchillo Negro district, with an aggregate value of about $250,000 during the various times of production. The Chise area produced mostly fluorspar, although there are also a few small copper mines and prospects. In the southern Salado Mountains large siliceous tungsten-bearing fluorspar deposits have been explored but have not yet gone into production. The Hillsboro district in the Animas Hills is the largest and most productive district in the eastern mineral belt. Since its discovery in 1877, this district has produced more than $7 million worth of ore (Harley, 1934, p. 141), chiefly gold from fissure vein and placer deposits, with subordinate copper and silver, and minor lead, manganese, and vanadium. In addition, a large disseminated porphyry copper and molybdenum deposit with silver and gold has recently been explored at Copper Flats in the central part of this district, but it has not yet gone into major production. The Lake Valley district also produced several million dollars worth of ore during the 1880's, chiefly silver with minor lead from rich, shallow, oxidized replacement deposits in limestone (Harley, 1934, p. 141); about 50,000 tons of siliceous manganese ore was also produced from this district between 1942 and 1945 (Creasey and Granger, 1953). The Macho district, at the southern end of the eastern belt, has produced a few thousand tons of rich silver-lead-zinc ore from fissure veins cutting Tertiary volcanic rocks (Harley, 1934, p. 157).

The host rocks for the iron deposits at Iron Mountain include lower Paleozoic carbonate rocks, Lake Valley Limestone and carbonate beds in the lower part of the Magdalena Group, which have been metasomatically altered to garnet and locally hematite- and magnetite-rich tactite at and near the contact with several small granitic, rhyolitic, and monzonitic plutons of Oligocene age. In the Cuchillo Negro district to the southeast, most of the small, rich base- and precious-metal deposits are also in tactites developed along the contact of quartz monzonite stocks and smaller plutons and dikes of Oligocene age with Paleozoic carbonate rocks, chiefly limestone of the Magdalena Group. In the Chise area, large structurally controlled vein and manto replacement deposits of siliceous fluorspar formed in carbonate rocks in the upper part of the Magdalena Group and in the overlying Yeso Formation of Early Permian age. The fluorspar deposits in the Salado Mountains are also in Magdalena Group. The Animas Hills near Hillsboro consist of a breached domal uplift of Upper Cretaceous or lower Tertiary andesite surrounding a quartz monzonite stock of Laramide age which has been unroofed by erosion. Large, gold-rich veins radiate outward from the stock into the surrounding andesite. Rich placer gold deposits derived from these veins were developed in younger gravels to the north and east, and the stock in the core of the uplift contains a porphyry deposit of copper-molybdenum-silver-gold ore. To the north and south of the uplift, Fusselman Dolomite and El Paso Limestone, exposed by erosion, contain some small base metal-silver vein and replacement deposits, and also manganese and small vanadium deposits. The bonanza oxidized silver-lead and manganese deposits of the Lake Valley district are confined to a narrow zone near the middle beds of the Lake Valley Limestone. In the Macho district at the southern end of the eastern belt, Tertiary andesite is the host rock for fissure veins, some of which are adjacent to late Tertiary altered rhyolite or latite dikes. All the important deposits of the eastern belt, except those at Chise and Lake Valley, occur in the vicinity of felsic intrusive rocks of Late Cretaceous or early Tertiary age, and such an intrusive may occur at depth just south of Lake Valley.

Throughout much of its length, the western boundary of the eastern mineral belt is delineated by strong, high-angle fault zones, with the block to the west downthrown several thousand feet in places (pl. 1). The tactite deposits at Iron Mountain terminate against this northerly trending boundary fault and are also cut off on the east by a smaller parallel fault downthrown to the east. Although mineralization in the Cuchillo Negro district is more dispersed and structural control is not as pronounced, there are some small deposits in this district just east of this western boundary fault, and both north-northwesterly and easterly trending faults have localized mineralization in other parts of the district. Small tin deposits in the eastern part of the district lie along a large eastern boundary fault downthrown to the east (Heyl and others, 1983). The larger fluorspar deposits in the Chise area are controlled by a northerly trending shear zone, or occur along a major fault with the downthrown block to the east, about a mile and a half east of the western boundary fault; there is also minor copper and fluorspar mineralization along this boundary fault. Fluorspar-tungsten deposits in the southern Salado Mountains are confined to the upthrown northern block of an easterly trending normal fault, about a mile to the east of a strong regional northerly trending boundary fault downthrown to the west. Mineralization within this block is localized by smaller northeasterly trending fault veins and shear zones. The structural pattern in the Hillsboro district is more complex; here, the north-trending boundary fault on the western side of the district curves easterly for about a mile around the southern end of the district, before resuming its southerly course, and is intersected near its eastern bend by a strong northeasterly trending belt of veins and small faults extending across the district. In addition, the quartz monzonite stock of Laramide age in the center of the mineralized area is the focus of a swarm of dikes and veins which radiate outward from it in all directions; small discontinuous easterly trending faults also cut the lower Paleozoic sedimentary rocks, which are exposed both to the north and to the south of the andesite surrounding the Cretaceous stock, which is also of Laramide age. A big eastern boundary fault marks the eastern edge of the district. In the Lake Valley district, the richest ore bodies were developed on the northern, upthrown side of a strong northwesterly trending fault which cuts off the western boundary fault about a mile and a quarter to the northwest; this northwesterly trending fault is traceable to Tierra Blanca Creek, east of Tierra Blanca Mountain. In the Macho district, ore occurs at the intersection of north-trending silver-lead-zinc veins with a northeasterly trending belt of shears, dikes, and veins, which disappears under younger gravel about a half a mile to the northeast.

Wall-rock alteration in the eastern belt is strongly influenced both by the type of mineral deposit and the composition of the host rock. In the Iron Mountain district, impure sandy or shaley beds have been metasomatically altered to an iron-poor granulite composed of calcic pyroxene, potassium feldspar, plagioclase, quartz, and epidote; purer carbonate beds have been converted to an iron-rich tactite consisting of coarsely crystalline magnetite, fluorite, helvite, danalite, willemite, hematite, andradite, as well as many other calc-silicate minerals, adjacent to the granitic intrusive. A large number of other rare minerals also occur in this zone. This skarn grades outward through a zone of crenulated alternating layers of dominantly magnetite, and dominantly andradite garnet interbanded with danalite and associated with magnetite, willemite, helvite, and spinel (locally called ribbon rock), surrounded by an outer halo of recrystallized carbonate rock containing scattered inclusions of epidote and spinel. The ore deposits of the Cuchillo Negro district to the southeast are in a different skarn zone in limestone adjacent to quartz monzonite intrusives, characterized by a much simpler mineral assemblage of andradite garnet, magnetite, hematite, and epidote containing pockets of ore sulfides; farther from the contact the limestone has been replaced by jasperoid and fluorite along faults and fracture zones. Jasperoid is also closely associated with the fluorspar deposits of the Chise area to the south, both as large blanket replacement bodies, and narrow "reefs" along fault zones. This same type of siliceous alteration of carbonate host rock also characterizes the fluorspar deposits of the southern Salado Mountains. Alteration in the Hillsboro district is more varied and complex. The central quartz monzonite stock shows weak but pervasive sericitic alteration of feldspar and chloritic alteration of biotite, with disseminated pyritic alteration adjacent to veins and mineralized fractures. The surrounding andesite exhibits intense propylitic alteration adjacent to the vein walls; latite dikes, which commonly lie adjacent to these veins, are intensely sericitized, and also locally pyritically altered. Carbonate rocks, in the lower Paleozoic section exposed on the northern and southern flanks of the Hillsboro district, contain jasperoid pods and reefs along mineralized faults and fracture zones. Silicification was the dominant alteration in the Lake Valley district, where a layer of jasperoid forms the floor of the silver-manganese deposits in the central and northern parts. In the southern part of the district, close to the boundary fault, this floor has locally been brecciated and recemented by ore. Andesite along a branch of this big northwest-trending fault zone south of the district exhibits strong propylitic alteration, similar to that adjacent to the veins in andesite in the Hillsboro district. Veins in the Macho district at the southern end of the belt have a strongly zoned alteration. Adjacent to the vein wall, andesite has been converted to a mixture of hematite, magnetite, sericite, and barite in a quartz gangue for 2 or 3 ft from the vein; this grades outward into a strongly argillized zone with a light-yellowish color,

about 5 ft wide, where it changes abruptly to a zone of propylitic alteration which extends outward, with gradually diminishing intensity, to as much as 50 ft from the vein.

Most of the mineral deposits of the eastern mineral belt, like those of the western belt, are considered to be of middle to late Tertiary age, although those at Lake Valley may be as young as Quaternary, and those at Hillsboro are of Laramide age.

Geochemical samples collected from the eastern mineral belt include a siliceous iron tactite sample from Iron Mountain, and jasperoid samples from various localities in the Cuchillo Negro, Chise, south Salado Mountains, Hillsboro, Lake Valley, and Macho districts and mineralized areas, detrital magnetite samples from Iron Mountain, Cuchillo Negro, and Lake Valley areas, and nonmagnetic heavy-mineral concentrates from stream-sediment samples at various localities within the eastern belt from Iron Mountain southward to Lake Valley. Some samples of each of these media exhibited geochemical anomalies related to mineralization. These will be discussed in detail under the subheading "Geochemical Information" under the appropriate area headings in the following section on mineralization in the eastern belt.

Metal anomalies in various sample media, which are not attributable to the presence of known mineral deposits in the vicinity, and may therefore indicate new unexplored mineralized areas, were found in several places. These include the following localities: (1) just south of the center of the north boundary of sec. 24, T. 10 S., R. 8 W., about a half a mile east of Reilly Peak, where a heavy-mineral-concentrate sample found near the head of a small gulch contains lead, copper, tungsten, and molybdenum plus visible fluorite; (2) the NW¼ sec. 6, T. 11 S., R. 7 W., on the northern edge of the Cuchillo Negro district in a small gulch draining southwest into Schoolhouse Canyon, a heavy-mineral concentrate from this locality was found to contain lead, silver, copper, and molybdenum, in addition, a jasperoid sample collected near this locality contains visible fluorite; (3) base- and precious-metal-bearing samples along upper Jaralosa Canyon associated with an Oligocene rhyolitic intrusive; (4) the SE¼ sec. 28 and NE¼ sec. 33, T. 11 S., R. 7 W., where several heavy-mineral concentrates in the upper drainage basin of a stream draining eastward from Cuchillo Mountain contain lead and silver and where one sample also contains bismuth; (5) the southern half of sec. 25, T. 11 S., R. 7 W., along Willow Spring Draw, heavy-mineral concentrates show lead, zinc, and tin, and jasperoid samples contain much tin and beryllium and moderate amounts of arsenic, antimony, and tungsten; (6) NW¼ sec. 13, T. 13 S., R. 7 W., copper and lead in heavy-mineral concentrates from near the mouth of a tributary draining north into Palomas Creek from the

Garcia Peaks; (7) NE¼ sec. 12, T. 14 S., R. 7 W., heavy-mineral concentrates contain tungsten and bismuth, and jasperoid samples from a northeast-trending fault zone reef associated with fluorspar deposits contain anomalous gold, beryllium, and tungsten; (8) secs. 1, 2, and 3, T. 16 S., R. 7 W., south of Black Peak and north of Percha Creek where heavy-mineral concentrates from several gulches contain abundant copper, lead, silver, and bismuth, jasperoid samples from fault-zone reefs associated with gold veins in andesite, close to the contact with underlying Paleozoic sedimentary rocks, are strongly anomalous in silver, beryllium, bismuth, molybdenum, lead, tin, vanadium, and zinc, and moderately anomalous in copper and tungsten. Samples from a large, unprospected jasperiod body along an east-trending fault in dolomite just south of the andesite contact contain abundant silver, copper, lead, tin, and vanadium, and moderately abundant arsenic, tellurium, molybdenum, and zinc, and; (9) NE ¼ sec. 29, T. 18 S., R. 7 W., a quarter of a mile southwest of the boundary fault at the southwestern edge of the Lake Valley district, heavy-mineral concentrates from a stream along a fault zone in altered andesite contain 3,000 ppm silver and 1% lead; other concentrates from sediment in this stream, both magnetic and nonmagnetic, are high in lead. A sample of calcite from a vein cutting andesite a quarter of a mile farther west is high in lead, zinc, mercury, and tellurium.

MINERALIZATION IN THE EASTERN BELT

The eastern mineral belt extends north-south through Sierra County and northward for about 6 miles into Socorro County to the canyon of the Alamosa Creek at the northern end of the Cuchillo Negro Mountains (the "Monticello Box"). Altered rhyolite tuff in this area contains the beryllium mineral bertrandite. Approximately 2 miles to the southeast of this locality, a small vein containing silver, lead, and copper minerals cuts the Tertiary volcanic rocks (W. R. Griffitts, oral commun., 1985). However, since only reconnaissance geology and little geochemical sampling have been done in the northern extension of the eastern belt, the discussion in this report is confined to the mineralized areas farther south, in Sierra County.

Iron Mountain Mineralized Area

Location

The deposits at Iron Mountain are on the western flank of the north end of the Sierra Cuchillo in northern Sierra and southern Socorro Counties. They occupy portions of sec. 35, T. 9 S., R. 8 W., and secs. 2, 3, 10, and 11, T. 10 S., R. 8 W. The main area is about 3½ miles long north-south and ¼–½ mile wide east-west. Some additional small mines and prospects in skarn extend the area of mineralization about 2 miles to the south and southeast, covering about 4 square miles in all.

Production

Massive complex skarn iron deposits were opened by small pits and adits in this area early in the 20th century, and small tonnages were probably shipped. Later, during the 1940's, small quantities of beryllium, fluorspar, and tungsten, locally associated with the iron ore, were produced and several carloads were stockpiled. Some tungsten ore was shipped to mills at Winston. A little red-bed copper and uranium ore in the Abo Formation may have been produced from small mines to the south, such as the Invincible mine about 1½ miles south of the iron deposits. Although prospecting has continued into the 1980's, the total value of the ore produced from this area is unknown but probably did not exceed $100,000 dollars.

Geologic Setting

This area has a complex geologic history. Initial broad folding of Paleozoic sedimentary rocks in early Tertiary time was followed by volcanic activity with the extrusion of andesites and more felsic volcanic rocks, and by block faulting and eastward tilting of the ancestral Sierra Cuchillo range. This was accompanied by the intrusion of numerous stocks and dikes of quartz monzonite, largely to the south of the Iron Mountain area, but also extending northward into it. Renewed tectonic activity in the area was accompanied by the intrusion of rhyolite plugs and dikes. Oligocene faulting in the area accompanied the emplacement of monzonite, rhyolite porphyry, and granite intrusives at Iron Mountain and the mineralogically complex contact-metasomatic aureole in the Paleozoic sedimentary rocks that surround them. This was followed by renewed uplift and high-angle faulting in the late Tertiary or early Quaternary (Jahns, 1955b). The Iron Mountain dikes and plutons intrude the Magdalena Group and the underlying strata. Adjacent to these intrusives, sandy or shaley beds have been converted to a granulite consisting of calcic pyroxene, amphiboles, potassic feldspar, quartz, and epidote with accessory calc-silicate minerals and garnet with chlorite-, calcite-, and sericite-filling fractures. Carbonate beds in the contact zone are replaced by an extremely complex massive tactite composed of coarsely crystalline magnetite, hematite, helvite, danalite, and andradite garnet; this tactite contains a wide variety of accessory minerals including many calc-silicate minerals, as well as fluorite, apatite, diopside, hornblende, quartz, vesuvianite, hematite, spinel, piemontite, biotite, chlorite, ludwigite, scheelite, powellite, willemite, bertrandite, beryl, pyrite, sphalerite, galena, and many other rare minerals. This was the material that was mined for iron ore and has been isotopically dated as Oligocene in age, the same age as the intrusives. Massive tactite grades outward through a zone of "ribbon rock" consisting of alternating crenulated layers of magnetite and of fluorite mixed with a variety of minerals including sericite, biotite, chlorite, vesuvianite, danalite, helvite, garnet, spinel, quartz, adularia, and opal, which is cut by late veinlets filled with fluorite, calcite, heulandite, and stilbite. This zone grades outward into recrystallized limestone containing pods of epidote and tremolite.

The small red-bed copper-uranium mines and prospects in the Abo Formation to the south and southeast of this tactite zone are north of Riley Peak, generally in close proximity to Oligocene quartz monzonite dikes, and younger rhyolite and felsite dikes, any or all of which may be genetically related to the mineralization.

The main deposits at Iron Mountain are cut off by high-angle northerly trending normal faults, both to the east and west. The geologic information on these deposits is from reports by R. H. Jahns (1944, 1955b) and from recent unpublished work by A. V. Heyl and C. H. Maxwell.

Geochemical Information

Rock samples.—A dump sample from an iron mine near the north summit of Iron Mountain, about a half a mile south of the county line, consists of spinel and vesuvianite cut by large, irregular masses of hematite, magnetite, and ilmenite, with younger veinlets and pods of quartz, fluorite, and sericite. This sample contains high concentrations of beryllium (30 ppm), bismuth (100 ppm), cadmium (70 ppm), cobalt (10 ppm), copper (1,000 ppm), fluorine (4%), indium (100 ppm), molybdenum (20 ppm), lead (2,000 ppm), tin (500 ppm), tellurium (2 ppm), vanadium (20 ppm), tungsten (300 ppm), and zinc (2%). A sample of ferruginous jasperoid was collected from a skarn copper-lead-zinc prospect in Magdalena Group near the northeastern end of a small northeast-trending ridge near the south boundary of sec. 11, T. 10 S., R. 8 W. (about a half a mile west of the Invincible mine). The sample is representative of alteration in the Paleozoic rocks associated with base-metal deposits about a half a mile south of the Iron Mountain granitic intrusive complex and 1½ miles south of the first sample. The sample is a metasomatically altered and silicified limestone from close to the contact with large rhyolite porphyry and aplite dikes. It consists dominantly of quartz and iron oxides and contains concentrations of as much as 20 ppm bismuth, 15 ppm

cobalt, 700 ppm copper, 100 ppm tin, 70 ppm tellurium, 1,000 ppm strontium, and 3,000 ppm tungsten.

Stream-sediment concentrates.—Two types of stream-sediment concentrate samples have been collected from this area. One consists of magnetic particles separated from the stream sediments with a hand magnet, which is largely detrital magnetite; the other consists of nonmagnetic heavy-mineral concentrates. Magnetic concentrates were collected at irregular intervals along a small stream draining westerly from near the southern end of Iron Mountain in secs. 9 and 10, T. 10 S., R. 8 W. A sample from near the head of this stream near a prospect pit in metasomatically altered lower Paleozoic rocks close to a large northeasterly trending quartz monzonite dike contains >1% manganese, 1% zinc, 1000 ppm tungsten, 700 ppm molybdenum, 700 ppm tin, and 20 ppm bismuth. Other samples collected farther to the west contained consistently anomalous amounts of manganese, molybdenum, and tin, as well as sporadically high concentrations of beryllium, bismuth, lanthanum, tungsten, and zinc, for about a mile and a half downstream (Lovering and Hedal, 1987). A nonmagnetic heavy-mineral concentrate collected north of Piñon Spring, near the eastern boundary of section 3 on the central-western flank of Iron Mountain, contains >1,000 ppm of both tin and tungsten. A sample from Havill Canyon, east of the northern part of Iron Mountain about a quarter of a mile north of the county line, contains >2,000 ppm lead and >1,000 ppm copper. Other samples from Havill Canyon, one close to the county line and one about a half a mile upstream in Sierra County, each contain >1,000 ppm tungsten; and the northern one also contains >1,000 ppm tin (Alminas and others, 1975b); the source of the tungsten is probably scheelite, but the source of the tin is unknown. Samples from Rouse Canyon, northeast of the Invincible mine in the southern part of the mineralized area, contain more than 2,000 ppm lead, but less than 1,000 ppm copper, although the mine produced mostly copper. A sample from farther down Rouse Canyon, just south of the hill from which the jasperoid sample was collected, contains >1,000 ppm tungsten, as does a sample from the upper part of the stream north of Rouse Canyon from which the magnetic concentrate samples were collected.

Summary and evaluation of geochemical samples.—Although only a few geochemical samples have been collected from this area, they are sufficient to indicate anomalously high concentrations in both iron tactite and jasperoid samples not only of tungsten and copper, which have been mined in the area, but also of tin, bismuth, and other base metals. The heavy-mineral concentrates (both magnetic and nonmagnetic) from streams draining the Iron Mountain area also contain anomalies for tin and tungsten, and in streams to the east and south of this area lead in the nonmagnetic concentrates appears to be the best indicator of base-metal mineralization, although concentrates from the Iron Mountain area are locally anomalous in many other elements.

Reilly Canyon–Deep Well Canyon Gap

There appears to be a small gap or break between the mines and prospects in the vicinity of Rouse Canyon in the southern part of the Iron Mountain mineralized area and those north of Red Hill Canyon at the northern edge of the Cuchillo Negro District mineralized area. It extends approximately from Reilly Canyon (Edwards Draw) to the divide between Red Hill Canyon and Deep Well Canyon, a distance of about 2½ miles. The big western range-front boundary fault, which marks the western edge of the eastern belt, trends south-southeast in this area. Immediately east of it, in the northern part of the gap, is a large intrusive mass of fine-grained leucocratic syenite or aplite a little more than a mile long and about a half a mile wide in the widest part, centered on Reilly Peak and called the Reilly Peak stock. It is intrusive into Magdalena Group and underlying Paleozoic sedimentary rocks on the north, south, and east, with many apophyses and small satellitic bodies along its borders. To the east of the Reilly Peak stock, a relatively undisturbed section of north-northeasterly trending, easterly dipping upper Paleozoic rocks consisting of the Magdalena Group and the Abo and Yeso Formations, extends eastward for about a half a mile in the upper drainage basin of Deep Well Canyon, where it is covered unconformably by the overlying lower Tertiary andesitic volcanic rocks. About a half a mile south of the Reilly Peak stock, a large northerly trending dike of Oligocene rhyolite porphyry intrudes the Magdalena Group carbonate rocks close to the western boundary fault, with several smaller dikes and bodies of the same rock trending northeasterly across Deep Well Canyon farther to the east, and also some smaller northwesterly trending quartz monzonite dikes. Limestone in the Magdalena Group adjacent to the contacts with both types of dikes has locally been silicified to form large bodies of jasperoid. Several samples of these jasperoids were collected for study and analysis. They are largely light gray, yellowish gray, pale red, and grayish red, with dense aphanitic to very fine grained texture. Spectrographic analysis of the samples shows no anomalous concentrations of base or precious metals, although most of them do contain slightly anomalous amounts of beryllium. However, several heavy-mineral concentrates taken from various localities in the upper basin of Deep Well Canyon to the east of the Reilly Peak intrusive contain >2,000 ppm lead, and one also contains >1,000 ppm copper. The source of these metals is not known.

Cuchillo Negro District

Location

The Cuchillo Negro district mineralized area extends along the main range of the Sierra Cuchillo from the ridge south of Deep Well Canyon, and southward for approximately 7 miles to the southern flank of Cuchillo Mountain, just north of the boundary between T. 11 and 12 S.; the width of the district is between 1 and 2½ miles. Mineral deposits in this area are widely separated. With several isolated prospects and small mines, there really is no well-defined central mining district (fig. 6).

Production

Mines in this area produced mostly copper and lead with a little zinc, silver, and gold (from contact-metasomatic ore). Total production from the district has been about 1,500 tons of ore with a market value of about $250,000, although no exact figures are available. Mines were chiefly active in the late 19th and early 20th centuries, although there was some production from the Dictator mine in the 1940's and 1950's; there has been relatively little activity during the past 50 years. The main mines in this district were the Vindicator, the Black Knife group, the Dictator, the Covington, and the Confidence (Harley, 1934, p. 116–123). In addition to these contact-metasomatic deposits, there are large fluorite veins north of Red Hill and also about 5 miles east of Red Hill, and a mile north of Carrizo Peak. Southeast of these deposits, in the S¼ sec. 25, T. 11 S., R. 7 W. near Willow Spring Draw, there are some small tin deposits, and there are also small, scattered red-bed copper deposits in shale of the Abo Formation at various places in the eastern part of the district. Although none of these deposits have any recorded production, the fluorspar deposits would certainly be mineable under favorable market conditions.

Geologic Setting

The main Cuchillo Negro district lies within an upraised block of Paleozoic Magdalena Formation carbonate rocks between the major northerly trending western boundary fault and another large northerly trending boundary fault about 2½ miles to the east with the eastern side downthrown (fig. 6). The late Paleozoic rocks within the belt have been intruded by numerous dikes and plutons of middle Tertiary quartz monzonite and also by dikes and smaller stocks of late Tertiary rhyolite. Precambrian metarhyolite and meta-andesite and overlying Cambrian to Mississippian sedimentary rocks are locally exposed in the western part of the belt, adjacent to the western boundary fault. Most of the ore deposits are in tactite zones in carbonate rocks

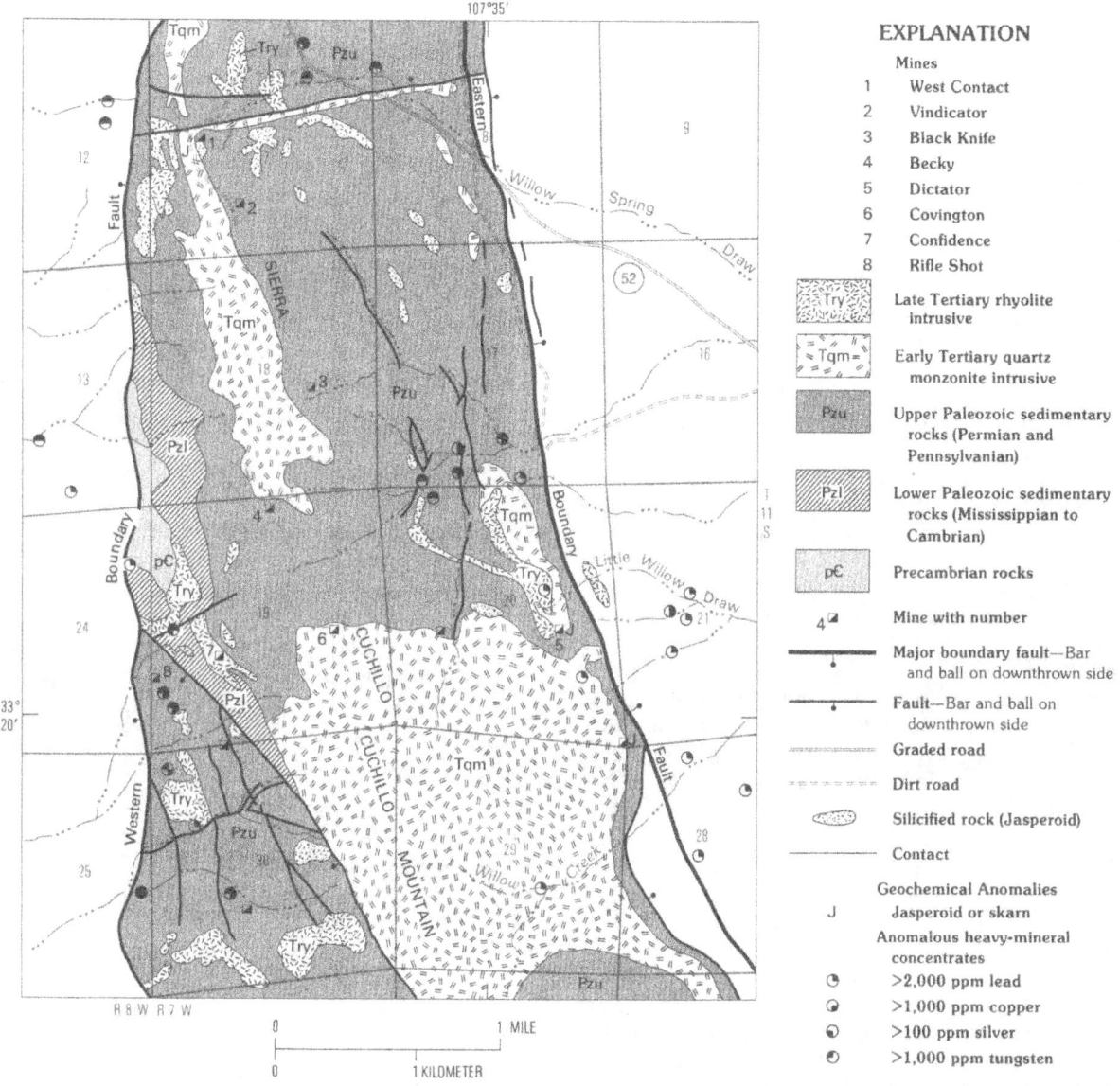

EXPLANATION

Mines
1	West Contact
2	Vindicator
3	Black Knife
4	Becky
5	Dictator
6	Covington
7	Confidence
8	Rifle Shot

Try — Late Tertiary rhyolite intrusive

Tqm — Early Tertiary quartz monzonite intrusive

Pzu — Upper Paleozoic sedimentary rocks (Permian and Pennsylvanian)

Pzl — Lower Paleozoic sedimentary rocks (Mississippian to Cambrian)

pC — Precambrian rocks

4 — Mine with number

Major boundary fault—Bar and ball on downthrown side

Fault—Bar and ball on downthrown side

Graded road

Dirt road

Silicified rock (Jasperoid)

Contact

Geochemical Anomalies
J   Jasperoid or skarn

Anomalous heavy-mineral concentrates
>2,000 ppm lead
>1,000 ppm copper
>100 ppm silver
>1,000 ppm tungsten

Figure 6. Sketch map of main Cuchillo district showing generalized location of geology, structural features (modified from A. V. Heyl, unpublished field map), mines, and anomalous geochemical sample localities.

of the Mississippian Lake Valley Formation or Pennsylvanian Magdalena Group, close to or along their contact with monzonitic or rhyolitic intrusives, where they form small vein, pipe, and podiform replacement ore bodies localized by tension fractures along the crests of small folds, or associated with dikes along fault zones. Contact-metasomatic alteration has produced a tactite consisting of magnetite, hematite, andradite garnet, and epidote, which is the host rock for small pockets and veins of pyrite, chalcopyrite, argentian galena, gold, and sphaler-

ite associated with a gangue of rhodochrosite, fluorite, quartz, and calcite. In the zone of oxidation, ore consists of cerussite, anglesite, azurite, malachite, smithsonite, acanthite, and minor cerargyrite associated with limonite and argentian manganese oxides. Outward from the contact, beyond the zone of metasomatic alteration, the limestone has locally been replaced by extensive reefs and large, irregular bodies of hydrothermal jasperoid and fluorite, many of which are structurally controlled by faults and fracture zones.

The fluorspar deposits near Red Hill, and to the east, are in large northerly trending fissure veins. Some of these contain only fluorite and calcite, others are quite siliceous. The tin deposits to the southeast are in close proximity to the large north-northwesterly trending eastern boundary fault (fig. 6); they contain cassiterite, hematite, and cristobalite in thin veinlets and breccia zones cutting rhyolite tuff of Oligocene age (Heyl and others, 1983).

Geochemical Information

Rock samples.—Jasperoid and tactite samples have been collected from several bodies in various parts of the district, mostly from outcrops along faults and fractures (Lovering, 1972). These are briefly described in approximate order of location from north to south and from west to east. (1) Samples from a northeast-trending fault zone reef cutting Magdalena Group on the ridge north of Red Hill Canyon near the southwest corner of sec. 30, T. 10 S., R. 7 W., are fine grained to aphanitic, dense, pale red, pink, and gray, and contain fluorite and as much as 150 ppm lithium. (2) Samples from a northeast-trending fault zone in limestone close to a small quartz monzonite stock on the hill south of Red Hill Canyon near the southeast corner of T. 10 S., R. 7 W., are dense, fine grained to aphanitic, grayish brown, grayish red, and dusky yellow, and contain abundant limonite and hematite; one sample contains vanadium (100 ppm) and molybdenum (150 ppm). (3) Samples from a small jasperoid body just south of the mouth of Red Hill Canyon near the western boundary fault in altered limestone near a small body of quartz monzonite are aphanitic, light to dark gray, and contain fluorite. (4) Several samples from an outcrop just east of the western boundary fault in the SW¼ sec. 6, T. 11 S., R. 7 W., in Magdalena Group are aphanitic, medium gray, brownish gray, and yellowish gray with fluorite and as much as 150 ppm lithium and 70 ppm vanadium; one sample also contains 200 ppm molybdenum. (5) Several samples from an outcrop of tactite near the west end of a strong easterly trending fault zone adjacent to the northern end of a quartz monzonite stock near the center of the western edge of sec. 7, T. 11 S., R. 7 W., are fine grained, grayish red, and yellowish orange, mostly hematite and limonite; samples contain as much as 150 ppm vanadium, 20 ppm bismuth, 1,000 ppm copper, 300 ppm tungsten, 1,000 ppm lead, 3,000 ppm zinc, and 5 ppm silver. (6) A sample from a large body of jasperoid in Magdalena Group just east of the western boundary fault near the southwest corner of sec. 7, T. 11 S., R. 7 W., consists of aphanitic, yellowish-gray fragments in a brownish-gray matrix; it contains fluorite, and 100 ppm vanadium, 200 ppm arsenic, 700 ppm barium, 3 ppm beryllium, 150 ppm lead, and 70 ppm tungsten. (7) A skarn sample from the dump of a small prospect at the contact between Magdalena Group and a quartz monzonite stock in the north-central part of sec. 18, T. 7 S., R. 11 W., is fine grained, light gray, and contains 100 ppm vanadium, 5,000 ppm lead, 7 ppm silver, 7 ppm bismuth, 20 ppm nickel, and 2,000 ppm zinc. (8) A jasperoid sample from a reef along a northerly trending fault in Magdalena Group 300 ft north of the Dictator mine in the central-eastern part of sec. 20, T. 11 S., R. 7 W., is dark gray, aphanitic, with 20 ppm vanadium and 30 ppm molybdenum. (9) A sample of tactite from the contact between Magdalena Group and a large quartz monzonite intrusive along a north-northwest-trending fault at a small mine in the southeast corner of sec. 20, T. 11 S., R. 7 W., is dense, fine grained, dark gray, contains galena, fluorite, and manganese oxides, and contains 2,000 ppm lead, 30 ppm silver, 700 ppm copper, 100 ppm vanadium, 5,000 ppm manganese, and 2.6% fluorine.

Stream-sediment concentrates.—Magnetic concentrates from a stream-sediment sample in the upper basin of Willow Creek in the SE¼ sec. , T. 11 S., R. 7 W., in a drainage basin entirely within the big Cuchillo Mountain quartz monzonite stock (fig. 6), are anomalously high in cobalt (70 ppm), chromium (3,000 ppm), and nickel (500 ppm) and also slightly anomalous in copper, lead, and zinc (Lovering and Hedal, 1987). A nonmagnetic heavy-mineral concentrate from the same locality showed 2% lead, despite the fact that there are no mines or prospects in the upstream drainage basin. Numerous compound metal anomalies are present in nonmagnetic heavy-mineral concentrates taken from both easterly and westerly trending stream drainages from the Cuchillo Negro district. Those streams draining westerly or southwesterly have a belt of lead anomalies >2,000 ppm extending from Deep Well Canyon in the north to the upper basin of Margarita Canyon in the south, a distance of about 7 miles; these samples also contain >1,000 ppm copper in the southern half of this belt. A few samples collected within a half a mile of mine workings in the upstream drainage basin show sporadic anomalies for silver, molybdenum, tungsten, and zinc. Streams draining eastward into Willow Spring Draw and upper South Well Canyon also contain abundant anomalous lead (>2,000 ppm) in the heavy-mineral concentrates in secs. 17, 20, 21, 22, 28, 33 and 34, T. 11 S., R. 7 W.; those in the southern half of sec. 17 also contain copper >1,000 ppm and molybdenum >70 ppm. Several samples in the SE¼ of sec. 28 contain >100 ppm silver, and one also contains >100 ppm bismuth (Alminas and others, 1975a, b, c).

Summary and evaluation of geochemical samples.—Tactite from the zone of metasomatic alteration in limestone bordering the quartz monzonite and rhyolite intrusives commonly contains high concentrations of a number of ore elements including arsenic, silver, barium, beryllium, bismuth, copper, lead, nickel, tungsten, vanadium, and zinc. Jasperoid from the abundant reefs and large, irregular replacement bodies in the area is generally barren of metals, although some bodies contain fluorite and lithium. The high chromium, cobalt, and nickel content of magnetic concentrates from the stream draining the big Cuchillo Mountain quartz monzonite stock is puzzling, as these elements are not normally enriched in accessory magnetite derived from so felsic an intrusive; however, they do concentrate in some magnetite-rich tactites in skarn zones bordering such intrusives. Thus, these magnetic concentrates may represent fossil placers derived from skarns that once overlaid this intrusive and that have been removed by erosion, or these magnetic concentrates may be derived from endomorphic zones within the quartz monzonite near the contact. The broad lead anomalies shown by the nonmagnetic concentrate samples both east and west of the district are disproportionate to the abundance and importance of lead in the known small, widely scattered ore bodies in the district; however, lead is commonly anomalous in tactite bodies surrounding intrusives, and pervasive mineralization in the tactites may be the source of the high lead values in the nonmagnetic concentrate samples.

South Cuchillo Mountain–Cuchillo Negro Creek Gap

An apparent gap in the mineralization of the eastern belt extends from the southern end of the Cuchillo Negro district mineralized area, approximately along the boundary between T. 11 S. and T. 12 S., southward to the valley of Cuchillo Negro Creek; the gap is approximately bounded on the west by Margarita Canyon and on the east by Montoya Canyon; the total area being about 2 miles north-south and 2½ miles east-west. This gap consists largely of upper Paleozoic sedimentary rocks of the Magdalena Group and the overlying Abo and Yeso Formations surrounding the southern border of the big Cuchillo Mountain quartz monzonite stock and another smaller stock of similar composition to the southwest between Margarita Canyon and Cañada Rancho de los Chivos. The smaller stock lies largely in the eastern half of section 6 and in the southwestern quarter of sec. 5, T. 12 S., R. 7 W. There is also a large, northerly trending dike of quartz monzonite close to the northeastern edge of this area, on the west side of Montoya Canyon, which intrudes limestone of the Magdalena Group. Structurally, the area is characterized by three major northerly trending fault zones: the western boundary fault runs southward along the east side of Margarita

Canyon, crosses the canyon about a mile upstream from the mouth, and goes through the town of Chise; the eastern boundary fault runs southward from the head of Montoya Canyon, down the eastern side of Montoya Canyon, and passes through Warm Springs on Cuchillo Negro Creek; and a third fault, parallel to and about midway between these two boundary faults, which follows the east side of Cañada Rancho de los Chivos creek northward from its confluence with Cuchillo Negro Creek east of Goat Camp Well continuing along the western flank of Cuchillo Mountain (fig. 6). The western and central fault zones are high-angle faults with the downthrown block to the west; the eastern boundary fault is downthrown on the eastern side. In addition to these major fault zones, the area is cut by a complex network of smaller faults and fracture zones, most of which trend northerly or northeasterly. Jasperoid samples were collected at three localities in this area. One sample from silicified Magdalena Group along the central north-trending fault zone just north of the mouth of Cañada Rancho de los Chivos is very fine grained, light gray and grayish red, and contains 50 ppm lithium. Several samples from a silicified sliver of limestone on the west side of the big eastern boundary fault in upper Montoya Canyon in the SE¼ sec. 33, T. 11 S., R. 7 W. are aphanitic, pale brown, olive brown, grayish red, grayish yellow or grayish orange, brecciated, with visible fluorite, and contain 50 ppm vanadium and 3 ppm beryllium. Silicified Tertiary dacite, andesite, and welded tuff samples from another locality a half a mile to the south along this same boundary fault in the NE¼ sec. 4, T. 12 S., R. 7 W. are fine grained, pale red and pale brown, and contain 30 ppm vanadium and 5 ppm beryllium, 5 ppm cobalt, and 30 ppm nickel.

Heavy-mineral concentrates from lower Margarita Canyon about a half a mile south of Apache Well contain >2,000 ppm lead and >1,000 ppm copper, possibly reflecting mineralization farther upstream in the southwestern part of the Cuchillo Negro district mineralized area. Another concentrate taken just above the mouth of Montoya Canyon to the east contains >1,000 ppm copper and >70 ppm molybdenum, possibly related to a small copper deposit prospected about a half a mile upstream from this locality. Several concentrates from the upper basin of the small unnamed creek which flows southward into Cuchillo Negro Creek about 1,000 ft west of Montoya Canyon contain visible fluorite, and a few of these also contain as much as >70 ppm molybdenum and >1,000 ppm tungsten, which may reflect mineralization related to northeasterly trending faults cutting the Magdalena Group in this area; there are no known mines or prospects in this drainage basin.

Chise Mineralized Area

Location

The main mineralization in the Chise area produced fluorspar deposits on the southern flank of Cross Mountain about 2 miles east of Chise, and in the hills to the south between Mine Road Canyon and Sophio Canyon in secs. 16 and 21, T. 12 S., R. 7 W. A few small widely scattered copper mines and prospects can be found just south of Chise in section 18 and about 3½ miles southeast of Chise near the central-western boundary of section 23.

Production

The only mines in the area with recorded production are the Victorio mine on the southern flank of Cross Mountain in the NW¼ section 16 which has produced about 70 tons of high-grade fluorspar and a few hundred tons of low-grade siliceous fluorspar ore, and the Chise fluorspar "mine" to the southeast of Cross Mountain. Subeconomic fluorspar deposits have also been prospected and drilled on the Mill Hill claim group about a mile to the south of the Victorio mine, as well as a small mine on the west side of Cross Mountain south of Cuchillo Negro Creek (McAnulty, 1978). The small Minta copper mine, just north of Madera Canyon near the central-western boundary of sec. 23, T. 12 S., R. 7 W., has no recorded production.

Geologic Setting

The fluorspar deposits are confined to a narrow northerly trending belt of Magdalena Group and the overlying Yeso Formation between the southern continuation of the major fault zone that follows Cañada Rancho de los Chivos and that of the eastern boundary fault that follows Montoya Canyon, with Tertiary volcanic rocks exposed both to the east and west of this belt in the southern part of the area, although the belt of upper Paleozoic rocks extends westerly from Cross Mountain to the western boundary fault in the vicinity of Chise. The high-grade fluorspar deposit at the Victorio mine is in a series of northerly trending, easterly dipping faults cutting Yeso Formation, about 50 ft apart and connected by a series of fluorite mantos as much as 6 ft thick. The deposit is about 2,600 ft long and as much as 500 ft wide in places, and the fluorspar ore is unusually pure. A large northerly trending stock of quartz monzonite forms the ridge and eastern flank of Twin Peaks with several satellitic bodies and dikes extending outward from its northern end into fine-grained, clastic sedimentary rocks of the Abo and Yeso Formations.

Farther south both the eastern and western flanks of this stock are covered by Tertiary volcanic rocks. Fluorspar deposits are localized within silicified zones along northerly and northeasterly trending faults, where they form pods, reefs, and small mantos of fluorite either alone, or in a gangue of quartz, or calcite, at various places through a stratigraphic interval of several hundred feet in the Magdalena and Yeso Formations. A very low-grade supergene copper deposit at the Minita mine is also localized along a small northeasterly trending fault zone that cuts older Tertiary volcanic rocks.

Geochemical Information

Rock samples.—Jasperoid was sampled from two localities in the Chise area, one of which is in a silicified dacite breccia along the eastern boundary fault about a half a mile northeast of the Victorio mine on Cross Mountain; the other on a north-trending fault zone in volcanic rocks on the Chise quadrangle boundary about 2,000 ft southeast of the Minita copper mine. Four samples from the first locality are dense, fine grained to aphanitic, varicolored in shades of light brown, reddish orange, and brownish, yellowish, and pinkish gray; they contain fluorite, limonite, and hematite, and as much as 30 ppm beryllium, 1,100 ppm fluorine, 70 ppm vanadium, 1,500 ppm barium, and 3,000 ppm titanium. Silicified andesite or dacite tuff from the second locality is dense aphanitic, medium gray, light olive gray, and brownish gray; it contains 30 ppm vanadium, 300 ppm arsenic, detectable (< 10 ppm) bismuth, and 100 ppm copper.

Stream-sediment concentrates.—Heavy-mineral concentrates from stream sediment contain visible fluorite at many localities in this area, specifically in the drainage basins of the two tributaries that flow northward into Cuchillo Negro Creek just east of Chise, at several localities along Mine Road Canyon south of Cross Mountain, along Sophio Canyon, and at the mouths of the two unnamed tributary gulches that empty into Cuchillo Negro Creek from the west, between Mine Road Canyon and Sophio Canyon. Fluorite was also found in concentrates from lower Madera Canyon below the Minita mine, and about a half a mile farther downstream. Concentrates containing metal are quite sparse. One concentrate from Cuchillo Negro Creek where it crosses the western boundary of section 9 (T. 12 S., R. 7 W.) contains >1,000 ppm tin and another panned about 300 ft downstream from this locality contains >2,000 ppm lead, as does one from the easterly draining tributary of Mine Road Canyon just below the Victorio fluorspar mine, and a second sample just above the mouth of this tributary. A heavy-mineral concentrate from Madera Canyon just below the Minita mine contains >1,000 ppm tin as well as visible fluorspar, but copper is not highly anomalous in this sample (Alminas and others, 1975a).

Summary and evaluation of geochemical samples.— Both jasperoid and pan-concentrate samples collected from the vicinity of major faults near the fluorspar deposits east of Chise contain visible fluorite, and heavy-mineral concentrates collected downdrainage from the principal faults in the vicinity of the fluorspar mines also show lead anomalies which probably reflect the common occurrence of galena as a minor accessory mineral in fluorspar deposits. Other metal anomalies are scarce in both media, suggesting that the Chise mineralized area is unlikely to contain important metal deposits; however, it is an important potential source of fluorspar.

Madera Canyon–Salado Mountains Gap

Another gap or break in the mineralization of the eastern belt, which contains no mines and few prospects, may be more apparent than real; it extends from the southern end of the Sierra Cuchillo, at approximately 33°15′ N lat., across the upper drainage basin of Madera Canyon, and southward about 8 miles to the Salado Mountains south of Salado Creek.

The big northerly trending boundary faults that delimit the northern part of the belt, all seem to die out and disappear in the northern part of this gap area. The big western boundary fault and the central fault seem to terminate about 3 miles south of Chise. The eastern boundary fault splits south of Cuchillo Negro Creek, with one branch continuing to several miles south of Palomas Creek, the other branch extending southeast to Palomas Creek north of Garcia Peaks. Another strong northerly trending fault zone, the Willow Springs fault which extends the entire length of the Chise quadrangle crossing Cuchillo Negro Creek and Madera Canyon about a quarter of a mile west of their junction, can be traced southward to the vicinity of Indian Canyon where it seems to end at the northern edge of the Double S Peak quartz monzonite stock (pl. 1). There is a discontinuous series of smaller faults to the south that are approximately in line with this major fault zone, but there does not appear to be a major structural break extending northerly through this entire area, although the western boundary fault of the Salado Mountains area extends north to just beyond Palomas Creek (pl. 1). The gap area is characterized by a plateau south of the southern foothills of the Sierra Cuchillo at Thumb Tank Peak and the ridge extending southerly from Twin Peaks, just west of Sophio Spring. This plateau consisting largely of Tertiary and Quaternary gravel, locally capped by flat mesas of Quaternary basalt, is crossed by the easterly

trending valleys of Cross O Canyon, the North and South Forks Palomas Creek, Avilas Canyon, and Salado Creek, which cut through an irregular and discontinuous northerly trending belt of hills at about the 107°30′ long. line. These include the Double S Peaks south of Madera Canyon between Indian and Board Gate Canyons, Yellow Mountain north of the confluence of Cross O Canyon and Palomas Creek, and the Garcia Peaks between Palomas Creek and Salado Creek. This line of hills is marked by several quartz monzonite stocks at and near Double S Peaks and in the Garcia Peaks area, which locally cut late Paleozoic sedimentary rocks and early Tertiary andesite. The big intrusive stocks of monzonite and quartz monzonite in this area do not appear to be associated with mineral deposits. There are a few small "red bed" copper prospects in Abo Formation southwest of Yellow Mountain in lower Quail Canyon and on a northerly flowing tributary of Palomas Creek about a mile east of Evans Ranch. This area is privately owned and access is restricted, consequently only recon-naissance geological mapping and geochemical sampling have been carried out; however, heavy-mineral concentrates from widely separated localities are variously enriched in tin, lead, and copper.

The only jasperoid sampled in this area came from a northerly trending reef in Paleozoic limestone about 3 miles southeast of Chise and a mile north of Thumb Tank Peak in sec. 29, T. 12 S., R. 7 W.; it is fine grained, dark gray and brownish gray, and contains neither fluorite nor anomalous amounts of any ore metals. Pan-concentrate samples collected near Sophio Spring and near the head of the north fork of Madera Canyon contain visible fluorite, as do several samples in tributaries of Cross O Canyon north of Grandview Mesa in the SW¼ section 3 (T. 13 S., R. 7 W.) in the NW¼ section 10, and also in Cross O Canyon west of Yellow Mountain. Some of these samples also contain >1,000 ppm tin as do samples east of Grandview Mesa in Quail Canyon both upstream and downstream from the road crossing in the SW¼ section 11; other samples also containing >1,000 ppm tin include from the vicinity of Evans Ranch near the con-fluence of the North and South Forks of Palomas Creek, from a short tributary gulch on the north side of Palomas Creek about ¼ mile northeast of Evans Ranch, and from downstream at the mouth of Cross O Canyon south of Yellow Mountain. One heavy-mineral-concentrate sample from a tributary of Board Gate Canyon close to the 5,565-ft benchmark in the SE¼ section 2 contains >2,000 ppm lead, and another from a north-draining tributary of Palomas Creek about a mile east of the Evans Ranch, downstream from a copper prospect, contains >2,000 ppm lead and >1,000 ppm copper. Thus, there is evidence of mineralization in the region between the Double S Peaks on the north and the Garcia Peaks on the south, which is also an area in which Permian sedimentary rocks are exposed.

Salado Mountains Mineralized Area

Location

The Salado Mountains are a small group of hills south of Salado Creek and east of Fox Canyon and Long Canyon, lying largely in the northeast corner of T. 14 S., R. 7 W., and covering an area about 3 miles long north-south and 1½ miles wide east-west.

Production

Although there is no recorded production from this area, several fluorspar deposits have been explored. The largest of these is approximately 1,000 ft long and 500 ft wide with an average thickness of 40 ft. It has been prospected and drilled, and contains an estimated 2.3 million tons of ore averaging 19.3% CaF_2. In addition, there are several smaller vein and replacement fluorspar deposits in the area which have not been explored or evaluated (McAnulty, 1978).

Geologic Setting

The Salado Mountains are an upraised block on the eastern side of a major northerly trending boundary fault along Fox and Long Canyons. Northward from the center of sec. 12, T. 14 S., R. 7 W., to Salado Creek, the Magdalena Group is exposed at the surface. It is cut off southward by an easterly trending fault with Precambrian granitic rocks exposed on the south side. Fluorspar mineralization is localized by a north-northeasterly trending fault zone of small displacement cutting through the central part of the limestone area and terminating at a small east-west fault. Several siliceous fluorspar vein and replacement deposits consisting of fluorite in a jasperoid quartz gangue are localized along this zone. Locally, large replacement manto deposits are cut by younger siliceous vein deposits.

Geochemical Information

Rock samples.—Jasperoid samples were collected from the large manto replacement deposit, and also from three localities along a north-northeast-trending fault that cuts the northern end of the manto. The manto jasperoid is dense fine grained to aphanitic, pale red to dark red, and contains hematite that formed contemporaneously with the jasperoid matrix. This manto is cut by veinlets of younger fluorite mixed with coarser quartz. Samples from along the north-northeasterly trending fault are heterogeneous in texture, with angular breccia fragments cemented by younger, locally vuggy matrix, and are mostly brown, ranging from light yellowish brown through grayish brown to very dark brown. The oldest generation of fine-grained quartz contains cubic limonite pseudomorphs after original pyrite; a younger generation of coarse vein quartz with fluorite followed, locally encrusted with goethite and manganese oxide; both generations are cut by veinlets of calcite. The manto jasperoid contains 1.5% fluorine, 3 ppm beryllium, and 20 ppm vanadium; jasperoid from the vein along the north-northeasterly trending fault contains more of these three elements (F to 27%, Be to 20 ppm, V to 200 ppm) and also contains as much as 70 ppm germanium, 100 ppm molybdenum, 10 ppm niobium, 1,000 ppm tungsten, and 50 ppm yttrium.

Stream-sediment concentrates.—Concentrates from lower Fox Canyon and the mouths of tributary gulches running north into Salado Creek from the Salado Mountains for a mile downstream contain abundant detrital fluorspar, and one sample from the mouth of a short gulch that joins Salado Creek near the boundary between the SE¼ and SW¼ of section 36 (T. 13 S., R. 7 W.) also contains >1,000 ppm of both tin and tungsten. Samples from the upper basin of a small easterly trending gulch near the west boundary of sec. 7, T. 14 S., R. 6 W., and in sec. 12, T. 14 S., R. 7 W., near the easterly trending fault between Magdalena Group to the north and Precambrian granite to the south, contain fluorite and >1,000 ppm tungsten; one sample also has >100 ppm bismuth. A sample from a small southwesterly draining tributary of Long Canyon near the center of the eastern boundary of sec. 11, T. 14 S., R. 7 W., contains >100 ppm bismuth, as does a sample from Long Canyon just south of the northern boundary of section 14 (Alminas and others, 1975b). Several samples from Long Canyon also contain fluorite.

Summary and evaluation of geochemical samples.— The abundance of fluorite in both jasperoid and heavy-mineral-sediment concentrates from this area reflects the known occurrence of fluorspar deposits in the area. However, the additional presence of high tungsten contents in both jasperoid and sediment concentrates, and of bismuth in a few of the sediment concentrates, suggests that this area also has potential for deposits of tungsten and possibly other metals.

Seco Creek–Tank Canyon Gap

Another apparent gap in the eastern mineral belt extends southward from the Salado Mountains to Tank Canyon just north of the Animas Hills, a distance of about 5 miles. The big boundary fault that follows Long Canyon dies out southward on the north side of Las Animas Creek about a half a mile upstream from the Ladder Ranch where it is cut off by a short southwesterly trending fault, that passes through the Ladder Ranch and connects with another fault that extends southeastward to the south of Ladder Ranch. The older Precambrian rocks that are exposed in the southern Salado Mountains to the east of the boundary fault are covered by Tertiary andesite east of the junction of Ash Canyon and Seco Creek. Another strong northerly trending fault with the eastern side upthrown, is traceable southward from Ash Canyon, about a half a mile upstream from the Artesia Windmill, crossing Las Animas Creek about ¾ mile below its junction with Cave Creek, thence across the western end of Bell Mountain, across the upper basin of Wanda Canyon, and across Tank Canyon near the boundary between secs. 15 and 16, T. 15 S., R. 7 W. A large fault extends southwesterly from the western boundary fault in lower Fox Canyon to upper Ash Canyon near its confluence with Big Rocky Canyon in NE¼ sec. 29, T. 14 S., R. 7 W., with the northwest side down (and apparently dies out about 1⅓ miles south of Ash Canyon). A strong southeast-trending fault extends southward from Las Animas Creek about a quarter of a mile west of Warm Spring past Wanda Tank and crosses Tank Canyon close to lat. 33°, with the downthrown side to the northeast (pl. 1). Upper Paleozoic sedimentary rocks are locally exposed in the upthrown block between these two faults, although they are covered in many places by extensive flows of Quaternary basalt, such as the one that caps Bell Mountain. Upper Paleozoic sedimentary rocks are also extensively exposed on the north side of Las Animas Creek from a point about a mile upstream from the Ladder Ranch, westward for about 2 miles to the northerly trending western boundary fault, and extending northward to the basalt flows that cap the divide south of Ash Canyon. To the east the Paleozoic rocks are covered by older Tertiary andesite, which also extends southward east of the southeast-trending fault.

Only a few heavy-mineral-concentrate samples have been collected from stream sediments in this area. One such sample from the mouth of an easterly draining gulch about a half a mile below the confluence of Ash Canyon and Seco Creek contains >1,000 ppm tin; another from the mouth of a small gulch running north from Bell Mountain into Las Animas Creek about a mile below its junction with Cave Creek contains >1,000 ppm copper. One from the upper drainage basin of Wanda Canyon on the south side of Bell Mountain near the center of section 10 (T. 15 S., R. 7 W.) contains >70 ppm molybdenum. Samples from short southerly trending tributaries of Tank Canyon near the eastern boundary of sections 15 and 14 (T. 15 S., R. 7 W.) contain >2,000 ppm lead, and one from another southeasterly trending gulch about a mile farther east has visible fluorite (Alminas and others, 1975a, b, c). None of these concentrates contain more than one ore element in large amounts, and there are no mines or large prospects in the area.

Hillsboro–Copper Flats District

Location

The old Hillsboro mining district extends from Tank Canyon on the north, southward about 5½ miles to Percha Creek, and from Warm Spring Canyon on the west eastward about 3.8 miles to the east side of the eastern boundary fault. The nearly circular area of the Animas Hills includes the recently developed disseminated porphyry copper-molybdenum-silver-gold deposits of Copper Flat in the central part, and covers approximately 20 square miles (pl. 1; fig. 7). Between the Animas Hills and the valley of Percha Creek to the south, there are several small silver, lead, zinc, vanadium, and manganese deposits in lower Paleozoic carbonate rocks, and a few similar deposits also occur to the north, between the Animas Hills and Tank Canyon.

Production

Since its discovery in 1877, the Hillsboro district has produced approximately $7 million worth of ore. This was dominantly gold, with considerable copper and silver, and with minor lead, manganese, and vanadium. About $2 million worth of gold was recovered from rich placer deposits in alluvial fans to the east of the hills, and the valley of Tank Canyon to the north. Most of the rest has come from several large mines on strong vertical fissure veins; these include the Rattlesnake (Snake), the Bonanza (Reward), the Richmond, the Opportunity, the El Oro–Andrews, the S. J. Macey vanadinite, and the Wicks mine claim groups in the hills surrounding the central porphyry stock which has undergone copper-molybdenum-gold-silver mineralization (Lindgren, Graton, and Gordon, 1910; Harley, 1934, p. 141). The Quintana mine has recently been developed in this stock as a large open pit but had not yet gone into production in 1987.

Geologic Setting

The host rocks for the most productive mines are andesite and latite flows, flow breccias, tuff, and volcanic agglomerate of Cretaceous and Paleocene age, which

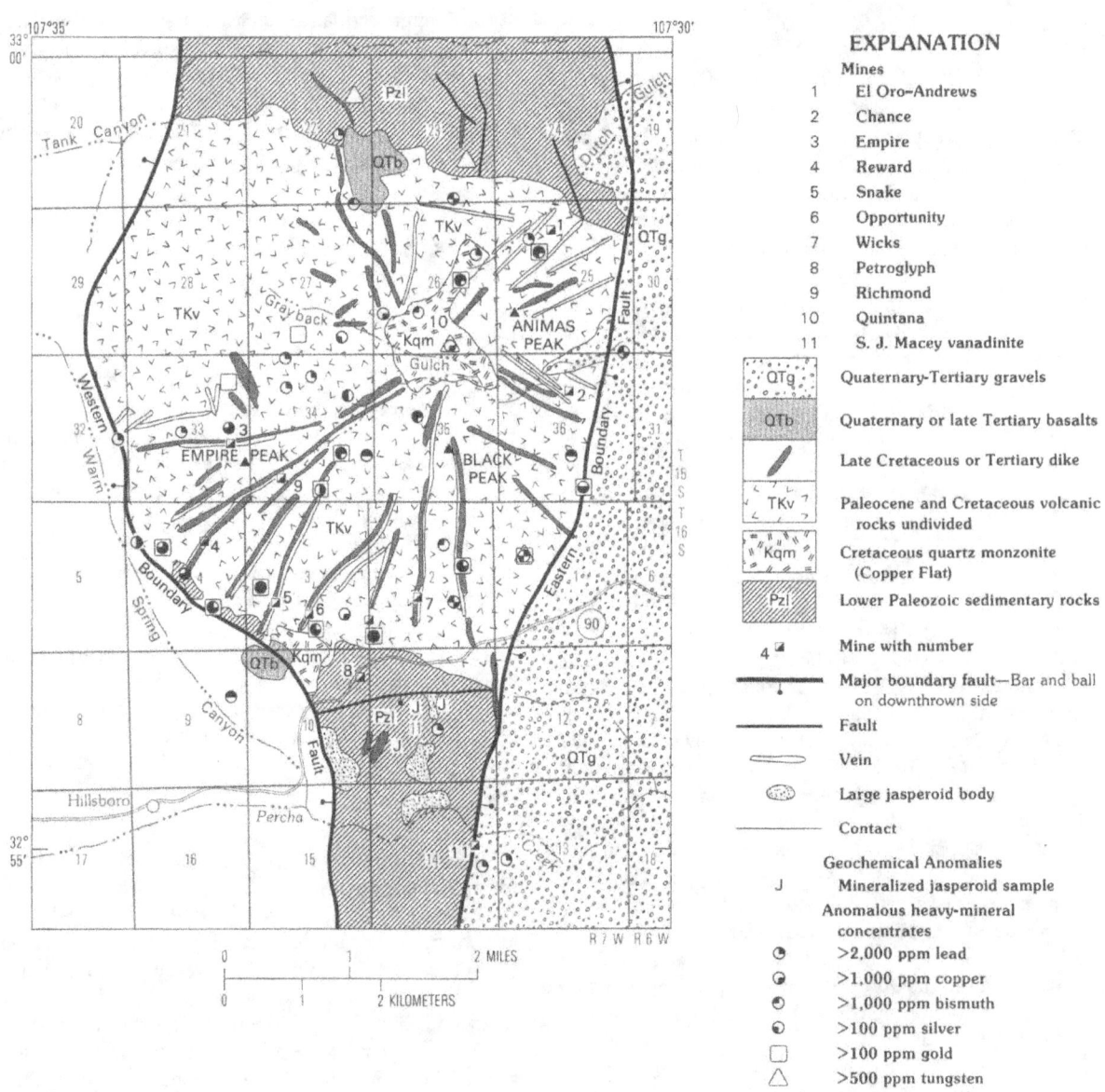

EXPLANATION

Mines

1	El Oro–Andrews
2	Chance
3	Empire
4	Reward
5	Snake
6	Opportunity
7	Wicks
8	Petroglyph
9	Richmond
10	Quintana
11	S. J. Macey vanadinite

QTg — Quaternary-Tertiary gravels

QTb — Quaternary or late Tertiary basalts

Late Cretaceous or Tertiary dike

TKv — Paleocene and Cretaceous volcanic rocks undivided

Kqm — Cretaceous quartz monzonite (Copper Flat)

Pzl — Lower Paleozoic sedimentary rocks

4 — Mine with number

Major boundary fault—Bar and ball on downthrown side

Fault

Vein

Large jasperoid body

Contact

Geochemical Anomalies

J — Mineralized jasperoid sample

Anomalous heavy-mineral concentrates

>2,000 ppm lead

>1,000 ppm copper

>1,000 ppm bismuth

>100 ppm silver

>100 ppm gold

>500 ppm tungsten

Figure 7. Generalized sketch map of the Hillsboro district showing general geology and structural features (modified from Hedlund, 1977a, and Harley, 1934), principal mines, and geochemical anomalies, with structural modifications by C.H. Maxwell.

make up the Animas Hills surrounding the mineralized Copper Flat quartz monzonite porphyry stock (fig. 7, Kqm), of the same Laramide age, which is exposed in the northeast-central part of the area. Lower Paleozoic carbonate rocks, consisting largely of El Paso Limestone and Fusselman Dolomite to the north and south of the area covered by Laramide volcanic rocks, are also host rocks for some small lead-zinc-silver-vanadium vein and replacement deposits, particularly to the south in the area between U.S. Highway 90 and Percha Creek. The

volcanic rocks are cut by radiating quartz veins and dikes. A northeasterly trending zone of shears and small faults has localized the emplacement of numerous latite dikes and quartz veins both northeast and southwest of the central stock. To the north of this zone in the western part of the mineralized area these veins and dikes trend nearly east-west; to the east of this zone in the southern part of the district they trend north-south, changing to a northwesterly trend in the southeastern part of the area. The big, nearly vertical boundary fault on the west side of the district, which is downdropped several hundred feet on the west side, changes to a southeasterly trend about 2 miles north of the mouth of Warm Spring Canyon, crossing Snake Gulch about ¾ mile above its mouth, and resuming its southerly trend across the junction of Warm Spring Canyon and Percha Creek. About a half a mile southeast of the point where this fault crosses Snake Gulch, it forms the western boundary of a small exposure of a much more mafic quartz monzonite body, which is apparently unmineralized and intrudes Paleozoic sedimentary rocks on its eastern border (fig. 7, secs. 3 and 10). Approximately a mile north of the junction of Warm Spring Canyon and Percha Creek, a strong easterly trending, nearly vertical fault, with the south side down, runs from the western boundary fault eastward for 1½ miles, at which point it is truncated by another north-trending boundary fault downthrown to the east. Where exposed along the Highway 90, a small outcrop of Precambrian gneiss and of Bliss Sandstone on the north side of the easterly trending fault is against Fusselman Dolomite on the south. Within the block of Paleozoic sedimentary rocks bounded by these three faults there are numerous small mines and prospects in the Fusselman Dolomite. Those mines in the eastern part of the block produced mostly vanadium and manganese; whereas, those in the central and western part contain more lead, silver, and molybdenum (Harley, 1934, p. 133–138). One of the more important vanadium minerals is endlichite, an arsenic-bearing lead vanadate.

The rich gold placers of the Hillsboro district, which produced approximately $2 million worth of gold, are concentrated along and just east of the large northerly trending boundary fault on the eastern side of the Hillsboro–Copper Flat district. The gold was concentrated in caliche-cemented zones in the upper part of the Pliocene and Pleistocene Palomas Formation. Two types of gold were found: coarse silver-bearing gold derived from the quartz veins, and fine "flour" gold, low in silver, derived from the Copper Flat porphyry stock. There may well be undiscovered placer gold deposits at the base of the Palomas Formation, where it rests on bedrock on the downthrown eastern side of the boundary fault. The thickness of the gravel adjacent to the fault remains unknown in 1987.

A small volcanic caldera, which is not shown on geologic maps of this area, is clearly visible from the Hillsboro-Kingston Highway 90 between the eastern and western mineral belts of this report. Its eastern rim lies about ¼ mile west of the high bridge over the Percha River west of Hillsboro, and its western rim is about 2½ miles farther to the west. Lake-bed sediments in the caldera are mostly of Oligocene age. No mineral deposits are known to be associated with this caldera, but it has

not been studied in detail; there could be some undiscovered deposits related to it (R.B. Taylor oral commun. to A.V. Heyl, 1978).

Wall-rock alteration in this district varies with the nature of the host rock. The Copper Flat quartz monzonite porphyry stock shows pervasive sericitic alteration of plagioclase and chloritic alteration of biotite, with pyritic alteration along veins and mineralized fractures. The andesite cut by the gold-bearing quartz veins shows intense propylitic alteration of the host rock adjacent to the vein wall. Porphyritic or andesitic latite dikes, which commonly form one wall of the veins, are sericitized and pyritically altered. In the Fusselman Dolomite, silicification in the form of small to very large jasperoid reefs, mantos, and irregular replacement masses, is the dominant type of hydrothermal alteration. Very large and prominent bodies of jasperoid are associated with the western boundary fault from Snake Gulch southward to Percha Creek, and also with the easterly trending cross fault. Smaller jasperoid bodies form reefs along fault veins and mantolike bodies capping hills in the Fusselman Dolomite between this cross fault and Percha Creek.

Primary ore minerals disseminated in the stock consist of chalcopyrite, bornite, and molybdenite with subordinate chalcocite associated with pyrite. The primary ore is unusually rich in copper and molybdenum. The zone of oxidation is very thin, with sulfides occurring within a foot of the surface in most places (Kuelmer, 1955). The supergene sulfide-enrichment zone, which is commonly less than 30 ft thick, contains silver, chalcopyrite, covellite, chalcocite, gold, chalcanthite, brochantite, and azurite. The quartz veins cutting the andesite contain free gold and auriferous pyrite associated with chalcopyrite, bornite, molybdenite, minor galena, and sphalerite in a gangue which locally contains a considerable amount of calcite in addition to quartz. Fault veins and replacement deposits in the carbonate rocks contain primary galena, sphalerite, and pyrite, which have generally been oxidized to a mixture of plumbojarosite, argentojarosite, cerussite, anglesite, willemite, hemimorphite, hydrozincite, and smithsonite with subordinate wulfenite, vanadinite, endlichite, mottramite, malachite, azurite, linarite, and descloizite in a gangue of quartz, calcite, limonite, and manganese oxides. Vanadinite and manganese oxides are particularly abundant in mines and prospects in the southeastern part of the area close to Percha Creek, and both were mined commercially.

Geochemical Information

Rock samples.—Jasperoid was sampled at five localities in the Hillsboro district. The first is a large mass of jasperoid capping a ridge along the western boundary fault on the east side of the highway, east of lower Warm

Spring Canyon; the second is a mass of silicified dolomite along the western boundary fault east of Snake Gulch near the southern end of the Rattlesnake vein; the third is an adit along the big east-west fault about midway between the two boundary faults; a fourth is a mineralized zone outcrop about 1,000 ft farther east along this fault; and the fifth is the dump of the Petroglyph mine in silicified limestone south of the southern end of the Ready Pay vein which cuts the andesite to the north, and is located about 0.4 miles northwest of the fourth locality. Samples from the first locality are aphanitic, brecciated, and locally vuggy in varying shades of red, yellowish brown, and yellowish orange, with disseminated limonite and hematite, and with chalcedony lining the vugs; samples contain as much as 5 ppm beryllium, 20 ppm vanadium, and 1,000 ppm zinc. Samples from the second locality are aphanitic, pale brown, pale orange, brownish gray, pinkish gray, and white; they contain pyrite altered to limonite, disseminated limonite, fluorite, and calcite, and contain as much as 50 ppm vanadium, 7 ppm beryllium, 1.2% fluorine, and lead (300 ppm) with detectable amounts of arsenic (200 ppm) and bismuth (<10 ppm). Samples from the prospect along the strong easterly trending fault consist of angular, aphanitic, gray inclusions in a fine-grained, locally friable and porous, yellowish-brown to moderate-brown matrix; they contain abundant limonite, and abundant beryllium (150 ppm), vanadium (150 ppm), and manganese (2,000 ppm); selected samples also contain as much as 3,000 ppm zinc, 500 ppm lead, 700 ppm arsenic, 300 ppm antimony, 150 ppm tungsten, 100 ppm nickel, 100 ppm yttrium, and 7 ppm silver. Samples from the mineralized outcrop farther east along this fault are even richer. They are fine to medium grained, locally vuggy, angular, gray inclusions in dark-reddish-brown to moderate-brown, vuggy matrices; and contain relict galena plus cerussite, anglesite, plumbo-jarosite, descloizite, jarosite, and limonite; they contain abundant lead (>10%), copper (300 ppm), zinc (1.5%), vanadium (700 ppm), tin (70 ppm), and silver (as much as 115 ppm), plus anomalous molybdenum (30 ppm), arsenic (1,500 ppm), tellurium (9 ppm), and a trace of gold (0.6 ppm). Two samples from the fifth locality, the dump of the Petroglyph mine, are really supergene ore samples in a yellowish-orange to brownish-gray, limonitic and jarositic jasperoid matrix, with relict galena and sphalerite altering to cerussite, anglesite, and willemite. These samples contain as much as 300 ppm silver, 15 ppm bismuth, 2% lead, 1,500 ppm vanadium, >1% zinc, beryllium (10–15 ppm), copper (200–300 ppm), mercury (10 ppm), molybdenum (500 ppm), tellurium (25 ppm), tin (10–15 ppm), and tungsten (50 ppm), plus 0.4 ppm gold.

Stream-sediment concentrates.—Heavy-mineral concentrates of sediment samples from streams and gulches draining the mineralized stock and the andesite hills containing gold veins commonly contain >2,000 ppm lead, >1,000 ppm copper; >100 ppm bismuth, >100 ppm gold, and >70 ppm molybdenum, and locally contain >1,000 ppm tungsten and >100 ppm silver within and adjacent to the mineralized area. Anomalies for lead and molybdenum generally appear to extend farther downstream than do those for copper, bismuth, and silver. A few samples from gulches draining the area of mineralized carbonate rocks at the northern and southern edges of the district contain >2,000 ppm lead (Alminas and others, 1977a, b, c; Watts and others, 1978a, b). Several concentrate samples from the apparently unmineralized northwestern part of the Animas Hills north of Grayback gulch and south of Tank Canyon (Watts, Alminas, and Kraxberger, 1978) contain visible fluorite, but lack metal anomalies (fig. 7).

Summary and evaluation of geochemical samples.—Neither of the jasperoid sample localities along the big western boundary fault yielded samples that were more than moderately anomalous in a few elements, such as beryllium and vanadium. However, both of the jasperoid localities along the strong easterly trending fault zone yielded more highly mineralized samples that contain anomalous concentrations of an extensive suite of ore elements, including silver, tin, arsenic, and tellurium, and the ore samples from the dump of the Petroglyph mine, which is at the contact between Paleozoic limestone and Laramide andesite on strike of a north-trending gold-bearing vein in the andesite, were richest of all, containing anomalous bismuth, mercury, gold, and tungsten in addition to the same anomalous elements found in samples from the east-west fault zone, suggesting the possibility of concealed multielement ore deposits in this area. The widespread occurrence of lead and bismuth in heavy-mineral concentrates from streams and gulches draining the mineralized area of the stock and surrounding andesite hills is also surprising, since neither of these elements is abundant in the ore deposits of this area.

Percha Creek–Berenda Creek Gap

From the southern edge of the Hillsboro district, just south of Percha Creek to the northern end of the Lake Valley district, about 2 miles south of Berenda Creek[3], there appears to be another gap in the eastern mineral belt about 12 miles wide. A low range of hills, called Sibley Mountain, extends southward from Percha Creek about a mile east of Hillsboro to just north of Berenda Creek, along the east side of this gap. This gap

---

[3]See footnote 1 on page 20.

is cut by easterly flowing streams of which Trujillo Canyon and Tierra Blanca Creek are the largest. The west side of the gap is covered with a broad belt of Quaternary gravel. From Percha Creek southward about 2½ miles to the canyon of Trujillo Creek, Lake Valley and Magdalena Limestones are exposed in these hills in a belt about a mile wide on the west side of the southward continuation of the eastern boundary fault from the Hillsboro district. The big western boundary fault is largely concealed beneath Quaternary alluvium south of Percha Creek. The eastern boundary fault runs into, and is truncated by, the western boundary fault just south of Trujillo Canyon. The western boundary fault changes to a southwesterly strike near Tierra Blanca Creek and appears to die out near the eastern end of Berenda Mountain. From Trujillo Canyon southward about 1½ miles to Oak Spring Creek, quartz latite flows and flow breccias of Oligocene age cover Sibley Mountain. Between Oak Spring Creek and Tierra Blanca Creek about a mile to the south, a large sill of older Tertiary andesite porphyry is exposed. Remnants of Magdalena Group overlie the intruding sill, which intruded along the western margin of the hills. From Tierra Blanca Creek to north of Berenda Creek, the hills are capped by quartz latite flows of Oligocene age and by overlying ash flow and airfall tuffs which are also considered to be of Oligocene age (Hedlund, 1977a). From near McGregor Ranch, the big Berenda fault extends northeastward along Berenda Creek which it crosses several times 1½ miles northwest of the Wilson Ranch, and continues to Sibley Gap, parallel to the southern portion of the western boundary fault, with which it is connected by a series of northerly trending cross faults (pl. 1). The southeastern side of the Berenda fault is upthrown. On the eastern side of the Berenda fault, southward from Berenda Creek, there are extensive exposures of Fusselman Dolomite, and south of the Wilson Ranch the overlying Percha Shale and Lake Valley Limestone are also exposed to the east of the Fusselman Dolomite, disappearing eastward beneath Quaternary gravel and Oligocene volcanic rocks.

There are no mines or prospects within this gap area, and the scattered exposures of Paleozoic carbonate rocks in Sibley Mountain and to the south of the E Nunn Ranch appear to be unmineralized. No visible fluorite or strongly anomalous concentrations of ore metals were found in heavy-mineral concentrates from stream sediments collected in this area. However, a few mineralized outcrops have been sampled (Watts, Alminas, and Kraxberger, 1978). The richest of these was limonitic and manganiferous clay from the contact between Percha Shale and the underlying Fusselman Dolomite collected on a ridge about 0.8 miles south of the Percha Creek canyon (sec. 23, T. 16 S., R. 7 W.) in the northern part of the area between the boundary

faults; it contains 3,000 ppm lead, 1,500 ppm zinc, 300 ppm each of arsenic, tin, and tungsten, 30 ppm molybdenum, and 15 ppm beryllium. Pyrite, calcite, and limonite from a fracture filling in Magdalena Group close to its contact with the andesite porphyry sill, along Oak Spring Creek, contains 70 ppm molybdenum. About 200 ppm of both zinc and arsenic were found along the northeastern contact between this sill and the limestone in the same area.

Lake Valley District

Location

About 15 miles south of Hillsboro, the small but rich Lake Valley silver-manganese district extends from just west of the town of Lake Valley northward for about 1½ miles along the southeastern flank of Apache Hill in secs. 16, 21, and 28, T. 18 S., R. 7 W. (fig. 8).

Production

Since its discovery in 1878, this district has produced nearly 6 million ounces of silver, some lead, and about 50,000 tons of siliceous manganese oxide ore. The shallow, rich silver deposits were largely mined out by 1893 (Lindgren, Graton, and Gordon, 1910), and the district lay dormant until wartime mineral shortages produced a short-term market for the manganese oxide ore associated with the silver deposits. Manganese was produced between 1942 and 1945 (Creasey and Granger, 1953); since then the district has again been dormant, although there has been some exploration drilling in the area recently. Most of the ore is oxidized and has come from three groups of mines—the Grande mine claim group just west of Lake Valley was the richest, including the famous Bridal Chamber which yielded 2½ million ounces of silver from one small silver chloride ore body; the Bella group, about ¼ mile to the north produced about 1 million ounces of silver; and the small Apache group at the northern end of the district produced about a quarter of a million ounces (Harley, 1934, p. 178–179). Most of the manganese ore also came from the southern, or Grande, group of mines (Creasey and Granger, 1953). Although oxidized ore minerals, such as arsenian vanadinite, wulfenite, anglesite, and cerussite, were associated with the silver, and some lead was undoubtedly produced, the amount and value of the lead production is unknown.

Geologic Setting

The belt of Paleozoic sedimentary rocks exposed on the eastern side of the Berenda fault extends southward from Berenda Creek in the vicinity of the Wilson Ranch, for about 3 miles to the southern end of

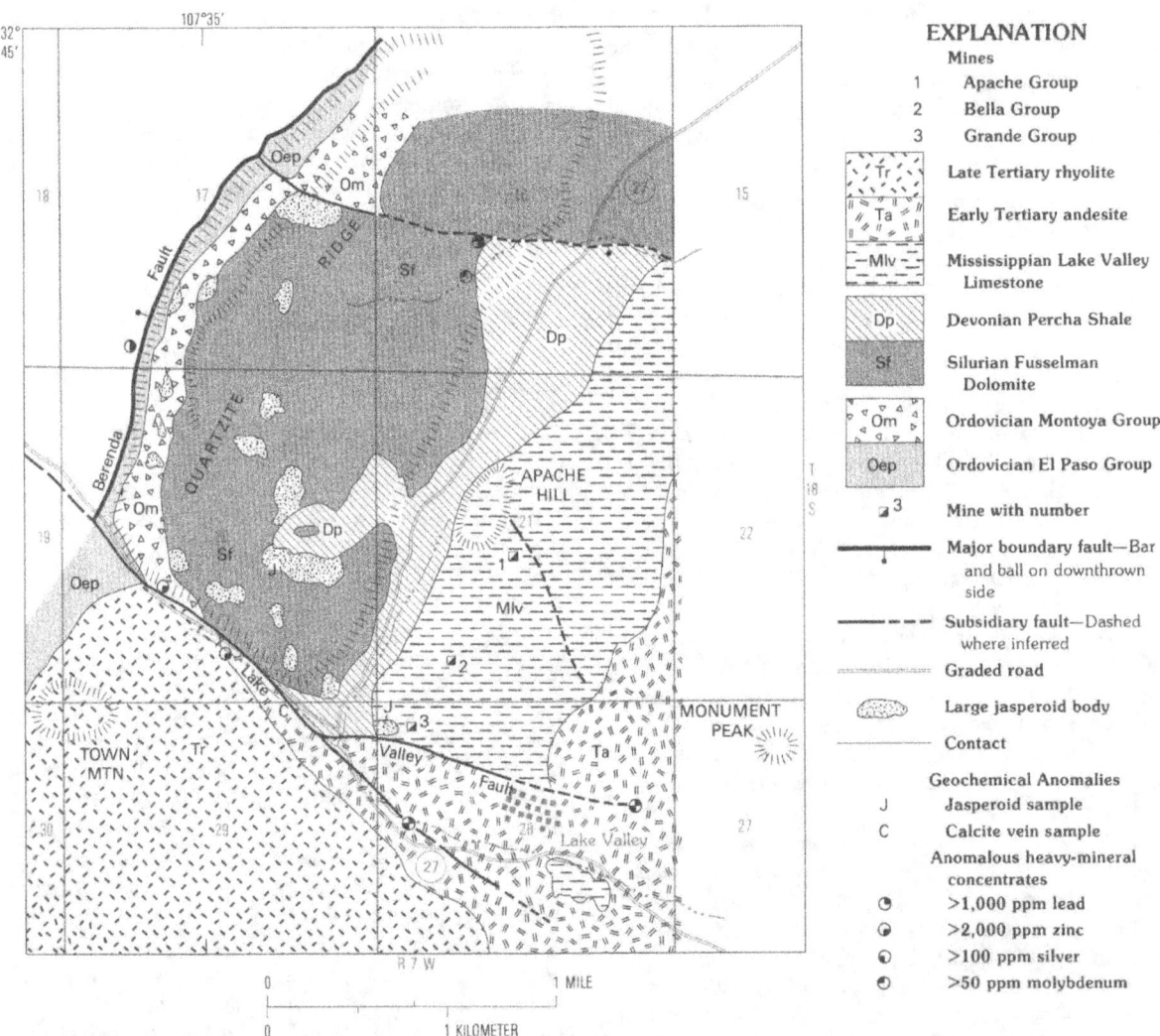

EXPLANATION

Mines
1 Apache Group
2 Bella Group
3 Grande Group

Tr Late Tertiary rhyolite

Ta Early Tertiary andesite

Mlv Mississippian Lake Valley Limestone

Dp Devonian Percha Shale

Sf Silurian Fusselman Dolomite

Om Ordovician Montoya Group

Oep Ordovician El Paso Group

3 Mine with number

Major boundary fault—Bar and ball on downthrown side

Subsidiary fault—Dashed where inferred

Graded road

Large jasperoid body

Contact

Geochemical Anomalies
J Jasperoid sample
C Calcite vein sample

Anomalous heavy-mineral concentrates
>1,000 ppm lead
>2,000 ppm zinc
>100 ppm silver
>50 ppm molybdenum

Figure 8. Sketch map of the Lake Valley District showing generalized geology (modified from Jicha, 1954, and Harley, 1934), and approximate location of structures, principal mine groups, and anomalous geochemical samples.

the Lake Valley district, where the Berenda fault terminates against the strong northwesterly trending Lake Valley fault, downdropped on the southwest side, which continues northwestward to Tierra Blanca Creek (pl. 1). On the northeastern side of this fault the belt of sedimentary rocks is about 2 miles wide with a gentle dip to the southeast. The western part of this section consists of El Paso Limestone and the overlying Fusselman Dolomite in a belt about a mile wide, making a broad cuesta called Quartzite Ridge. The Percha Shale which overlies the Fusselman Dolomite has been eroded to form a northeasterly trending valley which is followed by State Highway 27 northward. The overlying Lower Mississippian Lake Valley Limestone forms a parallel cuesta

on the southeast. This formation is subdivided into three members, the oldest of these called the Andrecito Limestone Member, overlies the Percha Shale and forms the crest of the ridge; the middle member, called the Alamagordo Limestone Member, is exposed on the upper eastern slope of the cuesta, with the upper or Nunn Limestone Member forming the lower slope. Silver ore bodies are localized at the contact between the Alamagordo and Nunn Members, largely in the basal 10 or 15 ft of the Nunn Member, but locally extending a few feet downward into the underlying Alamagordo Member. The overlying older Tertiary andesite south of the northwesterly trending Lake Valley fault is capped by a large mass of younger Tertiary rhyolite which forms

Town Mountain. Although the exposed contacts of this rock with the andesite are nearly flat lying, it may conceal a central intrusive vent. To the east of the Lake Valley district the older Tertiary andesites which overlie the Lake Valley Limestone are locally concealed beneath Quaternary alluvium. A large easterly trending normal fault about a mile and a half northwest of Lake Valley cuts and offsets the entire belt of Paleozoic sedimentary rocks east of the Berenda fault, and marks the northern limit of the mineralized area. Between this fault and the northwesterly trending Lake Valley fault to the south are some smaller northwesterly and northeasterly trending high-angle faults in the Lake Valley Limestone that do not extend westward into the Fusselman Dolomite, but these faults have localized ore deposits in the Lake Valley Limestone. These include the Columbia fault in the Bella mine group area and the Stone Cabin fault in the Apache mine group area farther north (Creasey and Granger, 1953; Harley, 1934, p. 172–176, 182–183).

Alteration in the carbonate rocks near ore was largely silicic. A layer of jasperoid 1–2 ft thick at the top of the Alamagordo Limestone Member in the northeastern part of the district forms the footwall of the silver ore bodies; the jasperoid layer thus separates the underlying manganese replacement deposits from the silver ore bodies. In the more strongly mineralized southern part of the district near the Lake Valley boundary fault, this jasperoid layer has been fractured, brecciated, and cemented by silver ore and younger jasperoid which locally extends downward a few feet into the Alamagordo Member. Manganese ore bodies in this area commonly replace the entire Alamagordo Limestone Member and locally extend downward into the upper part of the Andrecito Member beneath. Although this brecciated zone close to the boundary fault contains a few small reef-like bodies of argentiferous jasperoid, which is younger than the extensive jasperoid blanket at the top of the Alamagordo Limestone Member, the prominent jasperoid reefs along major faults, which are common in the southern part of the Hillsboro district, are notably absent at Lake Valley. In addition to the jasperoid blanket at the top of the Alamagordo Member previously described, there are numerous large, irregular replacement bodies of jasperoid in the Fusselman Dolomite and Montoya Group exposed along Quartzite Ridge, the large cuesta to the west; although some of these are quite close to the Lake Valley boundary fault (fig. 8), their forms are irregular and do not appear to be structurally controlled. Other similar bodies are abundantly exposed along this ridge for a mile or more to the north, and their presence probably is responsible for the name "Quartzite Ridge" given to this cuesta. The southern part of the district also has several calcareous travertine spring deposits of Quaternary or Holocene age along small faults and

fractures. Andesite along a subsidiary split of the big northwest-trending Lake Valley fault (fig. 8) shows intensive propylitic alteration associated with the fault zone. West of Berenda fault, in Ordovician rocks, are many jasperiod reefs, some mineralized.

Fault veins and fracture zones in the district locally contain varicolored argentiferous jasperoid, but more commonly they are filled with manganese oxide minerals, such as pyrolusite, ramsdellite, manganite, braunite, and psilomelane, associated with ankerite, limonite, and manganoan calcite. Sulfide minerals have been found only in the deepest mine workings, even there they are rare, and no cavities or their oxidized sulfide grain pseudomorphs exist in the oxide area of the main ore bodies. This supports the view that these are hypogene rather than supergene deposits. Black manganoan calcite and fine-grained quartz are the main gangue minerals. The bonanza replacement silver ore bodies in the Nunn Limestone Member consisted dominantly of cerussite and cerargyrite with sparser embolite, native silver, arsenian vanadinite, and wulfenite in a largely calcitic and limonitic gangue with relatively little manganese; whereas, the manganese replacement deposits in the underlying Alamagordo Limestone Member consisted dominantly of siliceous manganite and psilomelane with some braunite and cryptomelane associated with ankerite, manganoan calcite, limonite, and ferruginous jasperoid, with little silver.

Although these deposits are apparently quite young, Pleistocene or even Holocene in age, they were probably deposited from hydrothermal hot springs. Clark (1894, p. 165–167) thought that silver and manganese were originally disseminated in the Tertiary volcanic rocks that once covered the area, and the metals were leached by ground water during subsequent erosion to be preferentially concentrated at favorable sites in the underlying limestone. Harley (1934, p. 177) and Creasey and Granger (1953) attributed the deposits to hydrothermal solutions that rose along the Lake Valley and related subsidiary faults; the hydrothermal solutions precipitated ankerite, manganoan calcite, and according to Creasey and Granger, also deposited silver sulfosalts and argentiferous galena in veins and replacement deposits in the limestone. They thought that these minerals were subsequently remobilized and reconcentrated by ground water to form the oxidized ore bodies. However, Heyl and Maxwell (oral commun., 1980) regard most of the cerussite, silver halide minerals, lead vanadates and sulfates, manganoan calcite, manganese oxide minerals, and, possibly, the sparse galena as primary hydrothermal minerals derived from structurally controlled thermal spring waters, possibly related to those which formed the travertine deposits. Sulfides are extremely sparse in the deposits, no sulfide cavities occur, and the main minerals have all the textures of hot-spring deposits.

Geochemical Information

Rock samples.—Young and Lovering (1966) studied jasperoid from outcrops of some of the large, irregular replacement bodies in the Fusselman Dolomite and in the underlying Montoya Group as well as late argentiferous jasperoid from the Lake Valley Limestone collected from mine dumps just north of the Lake Valley fault. They found anomalously high concentrations of silver, lead, zinc, and molybdenum, plus other favorable characteristics, not only in the "ore" jasperoid samples from the mine dumps, but also in two of the bodies in the Fusselman Dolomite, with the most favorable body close to the Lake Valley fault and also near the contact with the overlying Percha Shale. This rock was brecciated, hematite stained, relatively coarse grained, and vuggy, and contained 30 ppm silver, 30 ppm molybdenum, 700 ppm lead, and 500 ppm zinc. Young and Lovering (1966) suggested that these favorable jasperoids might indicate the presence of hydrothermal replacement ore deposits in the Fusselman Dolomite beneath the Percha Shale and close to the Lake Valley fault. Vein calcite from the dump of a small prospect pit on the south side of the Lake Valley fault, about 500 ft southwest of the road junction in the valley between Quartzite Ridge and Apache Hill, contained about 25 ppm arsenic, 25 ppm lead, 130 ppm zinc, 0.04 ppm mercury, and 0.01 ppm tellurium, which are anomalous for vein calcite as a matrix.

Stream-sediment concentrates.—Two types of stream-sediment samples were collected in the area, nonmagnetic heavy-mineral concentrates and magnetic concentrates consisting largely of detrital magnetite along New Mexico State Highway 27, south and east of the mining district. Nonmagnetic heavy-mineral concentrates collected near Highway 27, northwest of the district from tributary gulches draining easterly from Quartzite Ridge, contain more than 2,000 ppm lead in the west-central part of section 16 where the big easterly trending fault crosses the valley; this sample also contains >10,000 ppm zinc and >70 ppm molybdenum. A sample from near the head of a northwesterly trending gulch at the south end of Quartzite Ridge along the Lake Valley fault also contains more than 10,000 ppm zinc (K. C. Watts, written commun., 1981). The gulch along the highway just south of the low hill of andesite capped by rhyolite south of the Lake Valley fault and about a quarter of a mile south of the Bridal Chamber mine (part of the Grande group) contains 1 lead and 3,000 ppm silver, directly over the trace of a southern split of the Lake Valley fault in NE¼ sec. 29, T. 18 S., R. 7 W. A lead- and silver-rich sample collected east of the mining district, near the cemetery, indicates that the area affected by mineralization is greater than the original mined area. Detrital magnetite concentrates collected at intervals along the road from the Lake Valley cemetery to a point about a quarter of a mile northwest of the road junction in the valley underlain by Percha Shale southwest of Apache Hill contain consistently anomalous concentrations of copper, lead, zinc, titanium, and niobium. The sample from the locality that yielded a lead- and silver-rich nonmagnetic concentrate contains 700 ppm lead and 50 ppm tin in the magnetic concentrate (Lovering and Hedal, 1987).

Summary and evaluation of geochemical samples.—The sporadic occurrence of favorable jasperoid in Fusselman Dolomite along the Quartzite Ridge west of the Lake Valley district, the high lead, zinc, and molybdenum contents of heavy-mineral concentrates from gulches draining this area, and the occurrence of lead, zinc, and silver in the gulch in volcanic rocks along the Lake Valley fault trace south of the Lake Valley district, all indicate that mineralization in the area was not confined to the known deposits in the Lake Valley Limestone but is more widespread both stratigraphically and geographically. In particular, the high lead and silver values along the southern split of the Lake Valley fault (fig. 8) separated by a ridge of volcanic rocks from the rich mines in the limestone on the north side of the Lake Valley fault, together with the strong propylitic alteration of the andesite along this fault, suggests leakage of metalliferous solutions along the fault zone from buried mineral deposits in a downfaulted block of Paleozoic carbonate rocks beneath the andesite between the branches of the fault.

Town Mountain–Macho Creek Gap

South of the northwesterly trending Lake Valley fault lies an apparently unmineralized gap that extends about 6 miles to the south side of Macho Creek in the SW¼ sec. 20, T. 19 S., R. 7 W. The large rhyolite-capped hill called Town Mountain lies at the north end of this gap. Only reconnaissance geologic mapping has been done in this hilly gap area, which is largely composed of older Tertiary andesite that is cut by a series of southeasterly trending stream drainages. No major faults have been recognized in this area. A few large northeasterly trending rhyolite and latite dikes cut the older andesite between Ricketson Draw and Macho Creek; these are shown on the geologic map of the Lake Valley quadrangle (Jicha, 1954a), and probably many smaller ones are not shown. No geochemical samples have been collected in this gap area.

Macho District

Location

The small isolated Macho district is about 6 miles south of Lake Valley and 1 mile north of Greg Hills, east of Macho Creek in the NE¼ sec. 30, in the NW¼ sec. 29, and in the SW¼ sec. 20, T. 19 S., R. 7 W. The main mineralized area trends northeasterly and is about a half a mile long by a quarter of a mile wide. This area, about 2 miles north of the Luna County line, appears to mark the southern end of the eastern mineral belt. The prospects and small shows of silver, lead, and zinc ores extend northward towards Macho Creek and also southward (oral commun., C. H. Maxwell to Heyl, 1984).

Production

This district produced approximately 3,000 tons of ore, averaging 20 lead and 15 ounces of silver per ton between its discovery in 1880 and 1928. Some zinc was also produced. The only producing mines in the district were the Old Dude and the adjoining Anniversary mine. These were, however, relatively deep mines with about 2,000 ft of workings on four levels, and two shafts 300 and 500 ft deep (Harley, 1934, p. 187–189; Jicha, 1954a, p. 74–77), in contrast to the mines of Lake Valley, the deepest of which only extended 150 ft beneath the surface.

Geologic Setting

The host rock for the deposits of the Macho district is the older Tertiary pyroxene andesite which is underlain at a depth of about 700 ft by Fusselman Dolomite. This underlying carbonate rock section has not been explored. The ore is in a system of northeasterly trending nearly vertical quartz veins from 1 to 3 ft wide, some of which are associated with highly altered parallel dikes of rhyolite or latite emanating from an intrusive center in the Greg Hills about a mile to the southwest. This vein and dike system is traceable for about 2 miles to the southwest, and nearly a mile to the northeast, where it disappears under alluvium, but the main mineralization appears to be restricted to the immediate vicinity of the mines. The veins exhibit an extensive and well-zoned alteration pattern. Adjacent to the vein walls is an intensely silicified zone, from 2 to 3 ft wide in which accessory ferromagnesian phenocrysts of the original andesite have been converted to hematite or magnetite with accessory sericite and barite in a quartz gangue. This grades outward into a belt of light-yellowish argillic or sericitic alteration about 5 ft wide, which is in sharp contact with an outer belt of greenish propylitic alteration that extends outward with gradually diminishing intensity for as much as 50 ft from the vein wall.

The veins have been oxidized to a depth of about 100 ft. Below this the primary ore minerals are argentian galena, chalcopyrite, and sphalerite with pyrite in a quartz gangue. In the oxidized zone, the ore consists of anglesite and cerussite with plumbojarosite, smithsonite, willemite, arsenian vanadinite, descloizite, and minor wulfenite, aurichalcite, and cerargyrite in a gangue of siliceous limonite, manganese oxides, quartz, and calcite. This supergene ore is largely porous and vuggy. Mineralization is mid-Tertiary in age, younger than the andesite, and penecontemporaneous with the rhyolite or latite dikes (Harley, 1934, p. 185–186).

Geochemical Information

The only geochemical samples collected from this area are some jasperoid samples from the dump of the Anniversary mine shaft. This jasperoid is dense, very fine grained to aphanitic in texture, and various shades of gray in color. Although it resembles the barren jasperoid from the Phillipsburg and Grafton districts in appearance, it differs markedly in composition. The complex mineralogy contains an extensive suite of hypogene and supergene ore and gangue minerals as tiny grains disseminated in the jasperoid quartz. These minerals include galena, sphalerite, chalcopyrite, pyrite, barite, ilmenite, sericite, and sphene as hypogene minerals, and anglesite, cerussite, hemimorphite, smithsonite, cerargyrite, kaolinite, limonite, calcite, manganese oxides, and leucoxene as supergene alteration products, plus several other minerals which could not be identified in thin section. Far from being barren, this jasperoid has an extensive suite of anomalous minor elements; some samples contain more than 2 lead, about 1 zinc, about 1,000 ppm of both copper and silver, about 0.5 ppm gold, and high concentrations of as much as 0.5 titanium, >0.5 barium, 1,300 ppm bismuth, 150 ppm cadmium, 0.2 strontium, 300 ppm vanadium, 500 ppm arsenic, 30 ppm cobalt, 4 ppm mercury, 50 ppm nickel, 1,000 ppm antimony, 1.6 ppm tellurium, and 50 ppm tungsten.

The large size, intensive alteration zones, complex mineralogy, and presence of calcite gangue in this vein system all appear to indicate that it may persist into the Fusselman Dolomite beneath the andesite cover. If so, there may be vein and replacement ore bodies in the carbonate rock. The chance for the existence of such deposits appears excellent.

OTHER MINERALIZED AREAS IN WESTERN SIERRA COUNTY

There are five mineralized areas in western Sierra County that lie outside the boundaries of the two mineral belts. These are: (1) the Taylor Creek tin deposits on the western side of the Black Range in the northwestern corner of the county; (2) the Terry uranophane-fluorspar prospect about 2 miles northeast of Monticello in the southwestern foothills of the San Mateo Mountains; (3) the Aragon Hill or Goldsborough gold district, just south of the Socorro County line in the southern San Mateo Mountains about 7 miles northeast of Monticello; (4) the gold prospects on Fifty-fifty Hill about 4 miles east of Monticello; and (5) the silver-lead-copper prospects in the Mud Springs Mountains northwest of Truth or Consequences (Hot Springs district).

Taylor Creek Tin Deposits

The Taylor Creek tin district, which occupies the northwestern corner of Sierra County (fig. 1), extends northward into Catron County and southwestward into Grant County. It is the most mineralized portion of a much larger arcuate belt of tin occurrences, which extends from west-central Grant County northward along the Black Range and then northwest and west nearly to the plains of San Augustine in western Catron County (C. H. Maxwell, oral commun., 1985). The portion of this district which is within Sierra County covers about 50 square miles in the upper drainage basins of Taylor Creek and Sawmill Canyon in T. 10 S., R. 10 and 11 W. Placer tin (cassiterite) was first discovered in the area in 1918 by J. N. Welch, a gold prospector. Tin mines are small and widely scattered, but there has been commercial production from at least one property near the head of Sawmill Canyon.

Cassiterite, accompanied by specular hematite, occurs in veinlets cutting soft, white, argillized rhyolite. Argillic alteration rarely extends for more than a foot outward from the veinlets (Harley, 1934, p. 67–72). C. H. Maxwell was making a detailed study of these deposits in 1985. The deposits appear to be unrelated to the base- and precious-metal deposits of the mineral belts (C. H. Maxwell, oral commun., 1985). No geochemical sample information on this area is available.

Terry Uranophane-Fluorspar Prospect

The Terry prospect is approximately 2 miles northeast of Monticello near the center of sec. 26, T. 10 S., R. 6 W., on the western side of the southern foothills of the San Mateo Mountains (Lovering, 1956, p. 368–371). The principal ore is fluorspar, and secondary uranium as the mineral uranophane, has also been found here, but not in sufficient quantity to be mineable for uranium. Although some fluorite has been mined, the amount and value of the fluorite production is not recorded. The deposit is in Lake Valley Limestone and the overlying basal Magdalena Group. The deposit is localized by the intersection of a northerly trending fault with a strong northwesterly trending regional fault zone.

The limestone at this fault intersection has been strongly silicified forming jasperoid reefs along the faults, and also large, irregular jasperoid mantos, replacing the carbonate rocks. Supergene silver, lead, zinc minerals are abundant.

Jasperoid samples were collected from three localities, the Terry prospect, a north-trending reef along a fault zone 500 ft east-northeast of the prospect, and a large, irregular jasperoid manto 900 ft east of the prospect in Magdalena Group. Jasperoid from the prospect is purple, gray, and red, banded and vuggy, and aphanitic to medium grained with disseminated grains of blue fluorite. It contains as much as 10% fluorine, 700 ppm vanadium, 700 ppm yttrium, 300 ppm lead, 1,000 ppm copper, 50 ppm tungsten, and 3,000 ppm barium; it also contains traces of silver (up to 10 ppm) and gold (up to 0.4 ppm). Jasperoid in the reef is brecciated, olive gray, grayish purple, white, and light to moderate brown, dense, and aphanitic to fine grained. It contains tellurium (±2 ppm), and some samples contain 150 ppm chromium, 200 ppm molybdenum, 100 ppm vanadium, and 5 ppm silver. Jasperoid of the manto is grayish red and light brown, generally dense and aphanitic. It is consistently anomalous only in vanadium (70 ppm).

The metals in jasperoid from the prospect and from the reef suggest leakage from structurally controlled vanadiferous lead-silver-copper molybdenum mineral deposits at depth.

Aragon Hill (Goldsborough) District

The small gold district of Aragon Hill is in the NE¼ sec. 6, and the NW¼ sec. 5, T. 10 S., R. 5 W., about 6 miles north-northwest of Monticello and just south of the Socorro County line. This district has produced an unknown, though probably small, amount of gold from small rich pockets of free gold in quartz-fluorite veins cutting an irregular rhyolitic intrusive body of late Tertiary age, which intrudes the older Tertiary andesite at Aragon Hill with an apophysis extending across the gulch to the next ridge to the east. The veins are localized in northerly trending shear zones from 3 to 10 ft wide cutting both the older andesite and the rhyolite intrusive, but the auriferous quartz veins are confined to the rhyolites. Ore consists of free gold, auriferous pyrite, fluorite, and limonite in a quartz gangue (Harley, 1934, p. 190–191).

No jasperoid is associated with the Aragon Hill gold quartz veins, but some jasperoid was collected about 1½ miles to the southwest in the SW¼ sec. 7, T. 10 S., R. 5 W., in an area where black continental shale and calcareous mudstone, interbedded with older Tertiary andesite and dacite flows are cut by northerly trending silicified zones. These samples are laminated in various shades of yellowish brown, grayish orange, dark red, and gray; they have an aphanitic to very fine grained texture and contain limonite pseudomorphs of cubic pyrite in addition to hematite and sericite in a matrix of chalcedony, opaline silica, and quartz. These samples contain no detectable gold (<0.1 ppm) but contained as much as 200 ppm niobium, 20 ppm tin, 100 ppm vanadium, and 200 ppm yttrium.

Fifty-Fifty Hill (Quartz Hill)

The prospects of Fifty-fifty Hill are just east of Questa Blanca Canyon in an isolated northerly trending ridge in the northwestern corner of sec. 8, T. 11 S., R. 5 W., just south of the southern San Mateo Mountains about 3½ miles east of Monticello. There are several small gold mines and prospects on the hill, but they have no recorded production (Harley, 1934, p. 192–193). The host rock is early Tertiary andesite and an overlying late Tertiary rhyolite cut by a north-trending fault zone through the middle of the hill with the rhyolite cap downthrown about 100 ft on the west side against the andesite on the east so the hill is now about half rhyolite and half andesite, hence the name "Fifty-fifty Hill." The west base of the hill is adjacent to another strong northwesterly trending fault zone whose western side is concealed beneath alluvium. Both the andesite and the rhyolite have been intensely silicified along the faults. The rhyolite cap on the west side, between the two faults now consists largely of silicified rhyolite breccia, in which most of the prospects are located. The ore in this silicified breccia zone consists of local low-grade disseminations of fine-grained free gold in a quartz-hematite gangue.

Jasperoid was sampled from brecciated rhyolite near the southwestern base of the hill, close to the large northwesterly trending fault, also from near the crest of the hill on both sides of the north-trending fault in white silicified rhyolite to the west and red silicified andesite to the east. At the southwest base of the hill, close to the northwest-trending fault, the jasperoid varies in color through various shades of brown, gray, and red; it is strongly brecciated, dense, and aphanitic to fine grained with abundant disseminated hematite and limonite. It contains from 1 to 2.5 ppm gold, up to 5 ppm tellurium, and up to 15 ppm silver, and also as much as 2,000 ppm barium and 5,000 ppm vanadium. One pale-brown sample from this locality also contains 20 ppm beryllium, 50 ppm cobalt, 1,500 ppm tungsten, and 200 ppm yttrium. Jasperoidized rhyolite from near the crest of the hill is largely white or light gray with a dense aphanitic texture, is anomalously high in beryllium (30 ppm) and niobium (50–70 ppm), but contains no detectable gold (<0.1 ppm). The silicified andesite to the east is generally pale red, pink, and orange, and contains as much as 50 ppm niobium, 3,000 ppm barium, 10 ppm

cobalt, 100 ppm lanthanum, 70 ppm vanadium, and 70 ppm yttrium; one sample contained 5 ppm tellurium. A calcite vein in the northwest-trending fault zone contains 6 ppm tellurium, 8,000 ppm manganese, 700 ppm barium, and 30 ppm vanadium; the residue from an acid leach of these calcite samples also contains 10 ppm beryllium and 50 ppm tungsten.

Mud Springs Mountains (Hot Springs District)

The Mud Springs Mountains are a low range of hills that extend from about 1 mile west of the northwest town boundary of Truth or Consequences northward for about 5 miles to Cuchillo Negro Creek. Several small mines and prospects are scattered at widely spaced intervals along the lower western flank of this range from its southern tip northward to its north end along Cuchillo Negro Creek. The total value of the ore from this area is unknown, but Harley stated that one of the mines near the south end of the range is reported to have produced about $40,000 worth of silver and lead ore (Harley, 1934, p. 193–194).

The range is essentially a long, easterly dipping cuesta of Precambrian granitic rocks, only exposed near the southwest end, overlain successively by Cambrian, Ordovician, Devonian, and Pennsylvanian sedimentary rocks in parallel belts to the east. The range is cut at intervals by several small faults, most of which trend northerly to northeasterly. The ore deposits are in veins and small replacement deposits in the Ordovician El Paso Limestone. The ore consists of galena, cerussite, copper carbonates, and supergene silver minerals with barite in a gangue of quartz, calcite, limonite, and manganese oxides. These rocks are locally cut by jasperoid reefs, which are most abundant near the contact between the Ordovician rocks and the unconformably overlying Magdalena Group. Jasperoid samples were collected for analysis at two localities in the area. One of these is in the northeast corner of sec. 31, T. 13 S., R. 4 W., near the south end of the range. The other is in the NE¼ sec. 11, T. 13 S., R. 5 W., near the north end. The first locality is an east-northeast-trending reef cutting El Paso Limestone. The jasperoid is moderate brown and dark reddish brown, locally vuggy, and fine grained to aphanitic; it contains as much as 2,000 ppm barium, 15 ppm beryllium, 500 ppm vanadium, 100 ppm tungsten, 50 ppm lanthanum, 1,000 ppm lead, 20 ppm yttrium, and 300 ppm zinc. The second locality is on a large northerly trending fault zone reef in the upper Montoya Group close to the contact with the overlying Magdalena Group. Jasperoid samples from this second locality consist of pale-pink and yellowish-gray to brownish-gray, dense, aphanitic breccia fragments in a locally vuggy, banded, dark-reddish-brown to dark-yellowish-brown and grayish-orange matrix. These samples contain up to 2,000 ppm barium, 30 ppm beryllium, 50 ppm lanthanum, 200 ppm lead, 500 ppm vanadium, 50 ppm tungsten, and 200 ppm zinc. The jasperoid at this locality terminates abruptly at a depth of about 15 ft, below which the fissure is filled with coarse-grained, vuggy, grayish-brown calcite. This calcite contains as much as 300 ppm arsenic, 1,000 ppm barium, 100 ppm vanadium, and 150 ppm zinc. The acid insoluble residue from the calcite also contains 50 ppm lead, 50 ppm beryllium, 700 ppm vanadium, and 1,000 ppm zinc.

SUGGESTIONS FOR PROSPECTING

Although most of the geochemical anomalies in the mineral belts are attributable either to the deposits of old mines in the vicinity, or to small ore mineral pockets in veins, there are a few localities in both the eastern and western mineral belts where geochemical and (or) geological evidence suggest the possibility of substantial concealed ore deposits. These areas include the following:

Western Mineral Belt

1. Upper basin of Chloride Creek—western Chloride district
2. Hoosier mine area—main Chloride district
3. Upper basin of Dark Canyon, south of Hermosa
4. Central Kingston district, north of Kingston
5. Southwest Canyon area, south of Kingston
6. Southern East Tierra Blanca district, east of junction of Cottonwood Creek and Pierce Canyon

Eastern Mineral Belt

1. Jaralosa Mountain
2. Southeastern Cuchillo Negro district, west of HOK Ranch
3. South-central Salado Mountains area
4. Southeastern Hillsboro district
5. Southern Lake Valley district
6. Macho district

These areas are discussed in detail below, with the location, the geological and geochemical evidence suggesting the presence of concealed deposits, and the probable nature of these deposits.

FAVORABLE AREAS FOR PROSPECTING IN THE WESTERN MINERAL BELT

Upper Basin of Chloride Creek

This area is located near the western boundary of sec. 20, T. 11 S., R. 9 W., where several quartz veins in andesite radiate outward from a center on Chloride Creek about ½ mile east of the Silver Monument mine shaft (fig. 1). These veins are from 1 to 8 ft wide with walls altered to narrow casings of pyritic and talc rock encased in a wider zone of propylitic alteration. Rich pockets of copper-silver ore consisting of bornite, covellite, chalcopyrite, and argentite, and their supergene alteration products, occur in vein-filling and replacement ore bodies along these veins, particularly in the Silver Monument mine, which has produced about $500,000 worth of copper and silver ore. The radial vein pattern suggests a buried intrusive in this vicinity. If such an intrusive exists, it could contain disseminated "porphyry" copper deposits, and also copper- and silver-bearing tactite ore bodies around its contacts.

Hoosier Mine Area, Chloride District

The Hoosier mine is about 1,000 ft east of the center of sec. 19, T. 11 S., R. 8 W., approximately 1¼ mile (2 km) west of the town of Chloride (fig. 3). This mine is on a strong northwesterly trending vein cutting the northern end of a large block of Magdalena Group which overlies Tertiary andesite. The mine produced about $15,000 worth of lead and silver ore from small replacement bodies of supergene cerussite, anglesite, and argentiferous manganese oxides in a gangue of talc and calcite. The geologic setting of this mine is very similar to that of the old St. Cloud mine to the south, where large and rich ore bodies of copper-silver ore were found where the vein cuts the underlying andesite beneath the Magdalena Group. However, the downward extension of the Hoosier vein, in the andesite, has yet to be explored.

Upper Basin of Dark Canyon, South of Hermosa

This area includes the unsurveyed NW¼ sec. 1, T. 14 S., R. 9 W., and the W½ of sec. 36, and most of the S 1/3 sec. 25, T. 13 S., R. 9 W., extending northward from the divide between North Seco Creek and Dark Canyon to the hills south and southwest of Flag Spring between the Seco Creek–Hermosa road on the west and the major eastern boundary fault on the east (fig. 4; pl. 1). Lower Paleozoic carbonate rocks are exposed in this area, and the hills are capped by Magdalena Group. A strong, locally mineralized fault zone extends northward from the divide up the east fork of Dark Canyon, which may be a southern continuation of a mineralized fault in the Hermosa district (fig. 4). A sample of jasperoid from a reef in the fault zone just south of the divide, close to the road, contains pyrite and shows high concentrations of silver, lead, zinc, and vanadium. In addition, a heavy-mineral concentrate from the mouth of a short westerly trending tributary of the east fork of Dark Canyon near the boundary between sections 25 and 36, close to this fault zone, contained >10,000 ppm zinc, >2,000 ppm lead, >100 ppm silver, and >70 ppm molybdenum. Small but rich silver-lead deposits in Lake Valley Limestone at the American Flag and Flagstaff mines, are located on a branch fault close to a north-trending fault about 1 mile north of this locality. There appears to be a good chance of other similar small deposits in or close to the strong north-trending fault in this area to the south of the American Flag mine.

Central Kingston District, North of Kingston

Between 1 and 2 miles north of Kingston, between Sawpit Canyon and Picket Spring Canyon, east of the junction of Sawpit and Picket Spring faults, is a large upraised isolated block of Precambrian granitic rocks intervening between the heavily mineralized area to the southwest and the north-trending belt of mines in Paleozoic sedimentary rocks to the east (fig. 5). This block contains no mines and few prospects. However, both Sawpit Creek at the northern end of this block and Picket Spring Canyon and its tributaries to the south of it yield heavy-mineral concentrates that contain large amounts of both molybdenum and tungsten close to this block of Precambrian granitic rocks. Although neither of these metals has been produced from the main district, tungsten mines in Precambrian granite do exist a few miles to the north. It is a strong possibility that vein or stockwork deposits of molybdenum or tungsten may be found in the granitic rocks close to the boundary faults.

Southwest Canyon Area, South of Kingston

About 1 mile south of Kingston, just north of the center of sec. 19, T. 16 S., R. 8 W, an isolated cluster of metal-rich heavy-mineral concentrates was found. A strong north-trending fault crosses the canyon at this place, with Magdalena Limestone on the west in contact with Tertiary rhyolite to the east. A heavy-mineral concentrate from the mouth of a small north-trending tributary gulch that follows the fault zone yielded >2,000 ppm lead, >1,000 ppm copper, >1,000 ppm tungsten, and >70 ppm molybdenum; and another sample from the mouth of the northeasterly trending gulch just upstream from this locality yielded >1,000 ppm tungsten and >70 ppm molybdenum. A third concentrate sample,

from a gulch that runs southerly into Southwest Canyon west of a small hill a few hundred feet farther upstream, contains >2,000 ppm lead (Alminas and others, 1977b, c; Watts and others, 1978a, b). This cluster of anomalies from an area containing no mines appears to be indicative of concealed base-metal and tungsten deposits.

South End of Tierra Blanca District—Eastern Part

This locality is just east of the junction of Cottonwood Canyon, and Pierce Canyon, about 2 miles south of the Silvertail mine, and 1¼ miles north of Signal Peak (pl. 1). Lower Paleozoic rocks consisting of Lake Valley Limestone, Percha Shale, and Fusselman Dolomite on the east side of Pierce Canyon are cut by a northwesterly trending fault to the south, and a northeasterly trending fault to the north. Heavy-mineral concentrates from stream sediments collected close to the fault junction yielded >2,000 ppm lead, >1,000 ppm tungsten, >100 ppm silver, and >70 ppm molybdenum (Alminas and others, 1977a, c; Watts and others, 1978a, b). This is the same suite of metals that were found in a heavy-mineral concentrate from a gulch just below the Silvertail mine. The combination of a favorable structural setting, favorable host rocks, and strong compound metal anomalies at this locality opposite the mouth of Cottonwood Canyon strongly suggests proximity of an undiscovered ore deposit.

FAVORABLE AREAS FOR PROSPECTING IN THE EASTERN MINERAL BELT

Iron Mountain Mineralized Area Including Jaralosa Mountain

Although the area around Jaralosa Mountain in the northern Sierra Cuchillo had not been mapped and sampled in detail prior to 1985, strongly altered rocks, and metals in stream-sediment concentrates, indicate that the area was mineralized. Volcanic rocks along Jaralosa Canyon west of Sullivan Spring show intense alunitic alteration. Heavy-mineral concentrates from the upper basin of Jaralosa Canyon at and east of Goat Spring contain more than 2,000 ppm lead, and a sample at Goat Spring also contains more than 1,000 ppm copper. Similar lead and copper values in heavy-mineral concentrates are also reported from several localities in the upper basin of Whitewater Canyon south of the county line and northeast of the summit of Jaralosa Mountain, where a large intrusive mass of rhyolite porphyry has been reported (Alminas and others, 1975b, c; Jahns, 1944).

Southeastern Cuchillo Negro District

This area is about 2 miles west of the HOK Ranch in the S½ sec. 28 and the NE¼ sec. 32, T. 11 S., R. 7 W., on the eastern flank of the Sierra Cuchillo (pl. 1). In this area a narrow belt of Magdalena Group lies between the eastern border of the big Cuchillo Mountain quartz monzonite stock and the regional eastern boundary fault, with Tertiary volcanic rocks east of the fault. Several heavy-mineral concentrates from the upper drainage basins of small streams draining easterly in this area are uniformly high in lead (>2,000 ppm). In addition, three samples near the eastern boundary of section 28 contain >100 ppm silver, and one of these also contains >100 ppm bismuth. A sample from a small stream in the NE¼ section 33 just west of the section line contains >2,000 ppm lead and >1,000 ppm copper (Alminas and others, 1975a, b, c). There are a few small prospects, but no mines, close to the boundary fault. The compound anomalies, particularly those for silver and bismuth, suggest proximity to a mineralized source which is probably a metal-bearing tactite in the Magdalena Group adjacent to the stock. Although a large deposit is unlikely in this area, there could be some rich copper-lead-silver ore bodies. Also, an unusually good area for fluorspar ore is at and near the large high-grade ore body just west of the mill site of the Chise fluorspar "mine" (pl. 1).

South-Central Salado Mountains Area

This locality covers about 1 square mile centered near the common corner of secs. 1 and 12, T. 14 S., R. 7 W., and secs. 6 and 7, T. 14 S., R. 6 W., in the Salado Mountains south of Salado Creek. This area is covered by Magdalena Group which is intruded by a small plug of quartz monzonite near the crest of the ridge about ½ mile to the west of the common section corner. There are some large siliceous fluorspar and jasperoid mantos to the south, northeast, and northwest of the corner, and a large vein with a northerly trend runs through the corner and continues northward for about 1 mile just east of the township boundary. Samples collected from a jasperoid reef along this vein, close to the section corner, contain visible pyrite and fluorite, and also sericite. Spectrographic analyses of these samples show high concentrations of tungsten (up to 1,000 ppm), and are also anomalous in molybdenum, vanadium, fluorine, beryllium, germanium, niobium, and yttrium. Heavy-mineral concentrates from stream sediments sampled in gulches both north and south of the area contain >1,000 ppm tungsten in addition to abundant fluorite, and a few of these also contain >100 ppm bismuth (Alminas and others, 1975a, b). The high tungsten content of both the jasperoid and the heavy-mineral concentrates from this area indicate a potential for concealed scheelite deposits.

Southeastern Hillsboro District

This area includes the W½ sec. 2 and the E½ sec. 3, T. 16 S., R. 7 W., south and southwest of Black Peak and north of U.S. Highway 90 (fig. 7). Hillsboro road. The area is covered by older Tertiary andesite in a broad belt between the Copper Flat stock to the north and Paleozoic sedimentary rocks which are exposed for a short distance north of the highway. A strong mineralized easterly trending fault zone cuts these sedimentary rocks just south of the highway. Several northerly trending gold veins, including the Ready Pay vein (between Wicks and Opportunity mines) and the Wicks vein, cut the andesite, southwest and southeast of Black Peak (fig. 7).

Jasperoid samples collected at two localities along the easterly trending mineralized fault south of the highway are rich in an extensive suite of metals, including gold, silver, copper, lead, zinc, molybdenum, tin, tungsten, nickel, arsenic, antimony, vanadium, beryllium, and tellurium. Siliceous ore samples from the dump of the Petroglyph mine contain highly anomalous silver, lead, bismuth, zinc, and vanadium, and also anomalous concentrations of copper, molybdenum, tin, tungsten, beryllium, mercury, and gold. This mine is in Paleozoic limestone just south of the contact with the overlying andesite and opposite the southern end of the Ready Pay gold vein which cuts the andesite to the north for about a mile. Heavy-mineral concentrates from southerly trending gulches in the andesite in this area show compound metal anomalies for gold, silver, copper, lead, and bismuth (Alminas and others, 1977a, b; Watts and others, 1978a, b, c).

These geochemical anomalies suggest a strong possibility that the buried contact of the mineralized Copper Flat stock with the Paleozoic carbonate rocks, beneath the andesite, may have localized skarn or other deposits of copper, tin, or tungsten. There is also a good possibility of major lead-zinc-silver replacement ore bodies in the limestone along the big north-trending veins in this area—comparable to those in the vicinity of the Hanover and Santa Rita stocks to the west (Hernon and Jones, 1968). The andesite cover over the Paleozoic rocks in this area probably has a maximum thickness of about 600 ft.

Southern Lake Valley District

West and southwest of the town of Lake Valley in the NW¼ sec. 28 and the adjacent northeast corner of sec. 29, T. 18 S., R. 7 W., is a favorable prospecting area (fig. 8). Just north of these parts of sections 28 and 29 are the shallow, rich silver ore bodies of the Grande group of mines that are in Lake Valley Limestone on the north side of the northwest-trending Lake Valley fault, which appears to have been a conduit for mineralizing solutions. On the south side of this fault, the Paleozoic carbonate section is concealed beneath early Tertiary andesite and overlain southward by a large cap of late Tertiary rhyolite. This south side area has never been drilled. A subsidiary split of the Lake Valley fault extends southeasterly along the gulch followed by the Lake Valley–Hillsboro road and along the northeastern flank of Town Mountain, farther west. The andesite along this fault zone exhibits intense propylitic alteration. Highly anomalous concentrations of silver (3,000 ppm) and lead (1%) were found in a heavy-mineral concentrate from a sediment sample taken from the gulch along the Hillsboro–Lake Valley road where it follows this fault zone. The gulch is separated from the mines of the Lake Valley district to the north by a low andesite and rhyolite-capped hill which lies between this fault and the main Lake Valley fault. Magnetic concentrate samples separated from this sediment sample and others along the gulch, both east and west of this locality, are consistently rich in copper, lead, and zinc, and one is also high in tin. A sample of vein calcite from a small prospect pit in the Lake Valley fault zone approximately ½ mile northwest of the Bridal Chamber mine (in Grande group) contains anomalously high concentrations of lead, zinc, mercury, and tellurium.

This evidence suggests a strong possibility of replacement lead-silver ore deposits in the downfaulted block of Paleozoic carbonate rocks underneath the Tertiary volcanic rocks between the two branches of the Lake Valley fault to the south and west of the old Lake Valley mining district.

Macho District

This area includes the Old Dude and Anniversary mines in the SW¼ sec. 20 and in the NE¼ sec. 29, T. 19 S., R. 7 W., about 1 mile north of the Greg Hills and ¾ mile west of Macho Creek. A strong northeasterly trending vein system associated with an intense alteration zone, and accompanied by parallel dikes of rhyolite or latite, cuts the older Tertiary andesite. The mines have produced lead-silver ore with some zinc from ore bodies in these veins to a depth of about 500 ft. Jasperoid samples from the dump of the Anniversary mine shaft contain a complex suite of hypogene and supergene ore and gangue minerals including galena, sphalerite, chalcopyrite, pyrite, barite, illmenite, hematite, sericite, and sphene as hypogene minerals, and anglesite, cerussite, hemimorphite, smithsonite, cerargyrite, kaolinite, manganese oxides, limonite, arsenian vanadinite, end-lichite, wulfenite, and leucoxene as supergene minerals, in a matrix of quartz and calcite. Chemical and spectro-graphic analyses of these samples yield an equally complex and extensive suite of anomalous ore metals and ore-indicator elements: lead, up to 2%; zinc, up to 1%; copper and silver, up to 1,000 ppm; gold, up to 0.5 ppm; plus highly anomalous concentrations of titanium, barium, bismuth, cadmium, strontium, and vanadium, and moderately anomalous amounts of arsenic, cobalt, nickel, mercury, tellurium, antimony, and tungsten. In fact, these samples are silver ore. The large size, intensity of alteration, and complex suite of ore and gangue minerals and minor element anomalies all indicate a vein system that extends to considerable depth. The abundance of calcite gangue in the veins suggests remo-bilization by hydrothermal fluids of calcium carbonate derived from Paleozoic limestone and dolomite beneath the andesite. This, in turn, suggests the possibility that vein and replacement base-metal-silver deposits are present in the carbonate rocks beneath the andesite. Although the total thickness of the andesite cover in this area is not known, it has been estimated to be in the neighborhood of 700 ft, or about 200 ft below the lowest mine workings in the district.

REFERENCES

Alminas, H. V., Watts, K. C., Griffitts, W. R., Siems, D. F., Kraxberger, V. E., and Curry, K. J., 1975a, Map showing anomalous distribution of tungsten, fluorite, and silver in stream-sediment concentrates from the Sierra Cuchillo–Animas uplifts and adjacent areas, southwestern New Mexico: U.S. Geological Survey Miscellaneous Investigations Map I–880, scale 1:48,000.

———— 1975b, Map showing anomalous distribution of lead, tin, and bismuth in stream-sediment concentrates from the Sierra Cuchillo-Animas uplifts and adjacent areas, southwestern Mew Mexico: U.S. Geological Survey Miscellaneous Investigations Map I–881, scale 1:48,000.

———— 1975c, Map showing anomalous distribution of molybdenum, copper, and zinc in stream-sediment concentrates from the Sierra Cuchillo-Animas uplifts and adjacent areas, southwestern New Mexico: U.S. Geological Survey Miscellaneous Investigations Map I–882, scale 1:48,000.

Alminas, H. V., Watts, K. C., Siems, D. F., and Kraxberger, V. E., 1977a, Map showing anomalous silver distribution in stream sediment concentrates, Hillsboro and San Lorenzo quadrangles exclusive of the Black Range Primitive Area, Sierra and Grant Counties, New Mexico: U.S. Geological Survey Miscellaneous Field Studies Map MF–900–C.

_____ 1977b, Map showing anomalous copper distribution in stream sediment concentrates, Hillsboro and San Lorenzo quadrangles exclusive of the Black Range Primitive Area, Sierra and Grant Counties, New Mexico: U.S. Geological Survey Miscellaneous Field Studies Map MF–900–D.

_____ 1977c, Map showing anomalous molybdenum distribution in stream sediment concentrates, Hillsboro and San Lorenzo quadrangles exclusive of the Black Range Primitive Area, Sierra and Grant Counties, New Mexico: U.S. Geological Survey Miscellaneous Field Studies Map MF–900–E.

_____ 1977d, Map showing anomalous zinc distribution in stream sediment concentrates, Hillsboro and San Lorenzo quadrangles exclusive of the Black Range Primitive Area, Sierra and Grant Counties, New Mexico: U.S. Geological Survey Miscellaneous Field Studies Map MF–900–F.

Clark, Ellis, 1894, The silver mines of Lake Valley, New Mexico: Transactions of the American Institute of Mining and Engineering, v. 24, p. 138–167.

Creasey, S. C., and Granger, A. E., 1953, Geologic map of the Lake Valley manganese district, Sierra County, New Mexico: U.S. Geological Survey Miscellaneous Investigations Studies Map MF–9, scale 1:24,000.

Dane, C. H., and Bachman, G. O., 1961, Preliminary geologic map of the southwestern part of New Mexico: U.S. Geological Survey Miscellaneous Investigations Map I–344, scale 1:360,000.

Harley, G. T., 1934, The geology and ore deposits of Sierra County, New Mexico: New Mexico School of Mines, New Mexico Bureau of Mines and Mineral Resources, Bulletin 10, 220 p.

Hedlund, D. C., 1977a, Geology of the Hillsboro and San Lorenzo quadrangles, New Mexico: U.S. Geological Survey Miscellaneous Investigations Map MF–900–A, scale 1:48,000.

_____ 1977b, Mineral resources map of the Hillsboro and San Lorenzo quadrangles, Grant and Sierra Counties, New Mexico: U.S. Geological Survey Miscellaneous Investigations Map MF–900–B, scale 1:48,000.

Hernon, R.M., and Jones, W.R., 1968, Ore deposits of the Central mining district, Grant County, New Mexico, in Ridge, J.D., ed., Ore deposits of the United States 1933–1967: New York, American Institute of Mining, Metallurgical and Petroleum Engineers, p. 1211–1237.

Heyl, A. V., Maxwell, C. H., and Davis, L. L., 1983, Geology and mineral deposits of the Priest Tank quadrangle, Sierra County, New Mexico: U.S. Geological Survey Miscellaneous Investigations Map MF–1665, scale 1:24,000.

Hill, R. S. 1946, Exploration of Gray Eagle, Grandview, and Royal John Claims, Grant and Sierra Counties, New Mexico: U.S. Bureau of Mines Investigative Report 3904, 7 p.

Jahns, R. H., 1944, Beryllium and tungsten deposits of the Iron Mountain district, Sierra and Socorro Counties, New Mexico: U.S. Geological Survey Bulletin 945–C, p. 45–79.

_____ 1955a, Possibilities for discovery of additional lead-silver ore in the Palomas Camp area of the Palomas (Hermosa) mining district, Sierra County, New Mexico: New Mexico Bureau of Mines and Mineral Resources Circular 33, 14 p.

_____ 1955b, Geology of the Sierra Cuchillo, New Mexico, in New Mexico Geological Society Guidebook of South-Central New Mexico: Sixth Field Conference, November 11–13, 1955, p. 159–174.

Jicha, H. L., Jr., 1954a, Geology and mineral deposits of Lake Valley quadrangle, Grant, Luna, and Sierra Counties, New Mexico: New Mexico Bureau of Mines and Mineral Resources Bulletin 37, 93 p.

_____ 1954b, Paragenesis of the ores of the Palomas (Hermosa) district, southwestern New Mexico: Economic Geology, v. 40, no. 7, p. 750–778.

Kuelmer, F. J., 1955, Geology of a disseminated copper deposit near Hillsboro, Sierra County, New Mexico: New Mexico Bureau of Mines and Mineral Resources Circular 34, 46 p.

Lindgren, Waldemar, Graton, L. C., and Gordon, C. H., 1910, The ore deposits of New Mexico: U.S. Geological Survey Professional Paper 68, 361 p.

Lovering, T. G., 1956, Radioactive deposits in New Mexico: U.S. Geological Survey Bulletin 1009–L, p. 315–387.

_____ 1972, Jasperoid in the United States—its characteristics, origin, and economic significance: U.S. Geological Survey Professional Paper 710, 164 p.

Lovering, T. G., and Hedal, J. A., 1987, Trace elements in magnetic concentrates from stream sediments in southwestern New Mexico—a potential tool for reconnaissance geochemical exploration in arid lands: U.S. Geological Survey Bulletin 1566, 31 p.

McAnulty, William, 1978, Fluorspar in New Mexico: New Mexico Bureau of Mines and Mineral Resources Memoir 34, 64 p.

Maxwell, C. H., and Heyl, A. V., 1976, Preliminary geologic map of the Winston quadrangle, Sierra County, New Mexico: U.S. Geological Survey Open-File Report 76–858.

_____ 1980, Mineralization and structure of mineral deposits in the Hermosa, Chloride, and Phillipsburg areas, New Mexico: Global Tectonics and Metallogeny, v. 1, no. 2, p. 129–133.

Watts, K. C., Alminas, H. V., and Kraxberger, V. E., 1978, Map showing areas of detrital fluorite and cassiterite, and the localities of rock samples, Hillsboro and San Lorenzo quadrangles exclusive of the Black Range primitive area, Sierra and Grant Counties, New Mexico: U.S. Geological Survey Miscellaneous Investigations Map MF–900–H, scale 1:48,000.

Watts, K. C., Alminas, H. V., Nishi, J. M., and Crim, W. C., 1978a, Map showing anomalous tungsten and gold in stream-sediment concentrates, Hillsboro and San Lorenzo quadrangles exclusive of the Black Range primitive area, Sierra and Grant Counties, New Mexico: U.S. Geological Survey Miscellaneous Investigations Map MF–900–I, scale 1:48,000.

_____1978b, Map showing anomalous lead in stream-sediment concentrates, Hillsboro and San Lorenzo quadrangles exclusive of the Black Range primitive area, Sierra and Grant Counties, New Mexico: U.S. Geological Survey Miscellaneous Investigations Map MF–900–J, scale 1:48,000.

_____1978c, Map showing anomalous bismuth distribution in stream-sediment concentrates, Hillsboro and San Lorenzo quadrangles exclusive of the Black Range primitive area, Sierra and Grant Counties, New Mexico: U.S. Geological Survey Miscellaneous Investigations Map MF–900–K, scale 1:48,000.

Young, E. J., and Lovering, T. G., 1966, Jasperoids of the Lake Valley mining district, New Mexico: U.S. Geological Survey Bulletin 1222–D, 27 p.

GOLD RUSH BOOKS

OREGON, USA

www.GoldMiningBooks.com

Books On Mining

Visit: www.goldminingbooks.com to order your copies or ask your favorite book seller to offer them.

Mining Books by Kerby Jackson

<u>Gold Dust: Stories From Oregon's Mining Years</u> - Oregon mining historian and prospector, Kerby Jackson, brings you a treasure trove of seventeen stories on Southern Oregon's rich history of gold prospecting, the prospectors and their discoveries, and the breathtaking areas they settled in and made homes. **5" X 8", 98 ppgs. Retail Price: $11.99**

<u>The Golden Trail: More Stories From Oregon's Mining Years</u> - In his follow-up to "Gold Dust: Stories of Oregon's Mining Years", this time around, Jackson brings us twelve tales from Oregon's Gold Rush, including the story about the first gold strike on Canyon Creek in Grant County, about the old timers who found gold by the pail full at the Victor Mine near Galice, how Iradel Bray discovered a rich ledge of gold on the Coquille River during the height of the Rogue River War, a tale of two elderly miners on the hunt for a lost mine in the Cascade Mountains, details about the discovery of the famous Armstrong Nugget and others. **5" X 8", 70 ppgs. Retail Price: $10.99**

Oregon Mining Books

<u>Geology and Mineral Resources of Josephine County, Oregon</u> - Unavailable since the 1970's, this important publication was originally compiled by the Oregon Department of Geology and Mineral Industries and includes important details on the economic geology and mineral resources of this important mining area in South Western Oregon. Included are notes on the history, geology and development of important mines, as well as insights into the mining of gold, copper, nickel, limestone, chromium and other minerals found in large quantities in Josephine County, Oregon. **8.5" X 11", 54 ppgs. Retail Price: $9.99**

<u>Mines and Prospects of the Mount Reuben Mining District</u> - Unavailable since 1947, this important publication was originally compiled by geologist Elton Youngberg of the Oregon Department of Geology and Mineral Industries and includes detailed descriptions, histories and the geology of the Mount Reuben Mining District in Josephine County, Oregon. Included are notes on the history, geology, development and assay statistics, as well as underground maps of all the major mines and prospects in the vicinity of this much neglected mining district. **8.5" X 11", 48 ppgs. Retail Price: $9.99**

<u>The Granite Mining District</u> - Notes on the history, geology and development of important mines in the well known Granite Mining District which is located in Grant County, Oregon. Some of the mines discussed include the Ajax, Blue Ribbon, Buffalo, Continental, Cougar-Independence, Magnolia, New York, Standard and the Tillicum. Also included are many rare maps pertaining to the mines in the area. **8.5" X 11", 48 ppgs. Retail Price: $9.99**

<u>Ore Deposits of the Takilma and Waldo Mining Districts of Josephine County, Oregon</u> - The Waldo and Takilma mining districts are most notable for the fact that the earliest large scale mining of placer gold and copper in Oregon took place in these two areas. Included are details about some of the earliest large gold mines in the state such as the Llano de Oro, High Gravel, Cameron, Platerica, Deep Gravel and others, as well as copper mines such as the famous Queen of Bronze mine, the Waldo, Lily and Cowboy mines. This volume also includes six maps and 20 original illustrations. **8.5" X 11", 74 ppgs. Retail Price: $9.99**

<u>Metal Mines of Douglas, Coos and Curry Counties, Oregon</u> - Oregon mining historian Kerby Jackson introduces us to a classic work on Oregon's mining history in this important re-issue of Bulletin 14C Volume 1, otherwise known as the Douglas, Coos & Curry Counties, Oregon Metal Mines Handbook. Unavailable since 1940, this important publication was originally compiled by the Oregon Department of Geology and Mineral Industries includes detailed descriptions, histories and the geology of over 250 metallic mineral mines and prospects in this rugged area of South West Oregon. **8.5" X 11", 158 ppgs. Retail Price: $19.99**

Metal Mines of Jackson County, Oregon - Unavailable since 1943, this important publication was originally compiled by the Oregon Department of Geology and Mineral Industries includes detailed descriptions, histories and the geology of over 450 metallic mineral mines and prospects in Jackson County, Oregon. Included are such famous gold mining areas as Gold Hill, Jacksonville, Sterling and the Upper Applegate. **8.5" X 11", 220 ppgs. Retail Price: $24.99**

Metal Mines of Josephine County, Oregon - Oregon mining historian Kerby Jackson introduces us to a classic work on Oregon's mining history in this important re-issue of Bulletin 14C, otherwise known as the Josephine County, Oregon Metal Mines Handbook. Unavailable since 1952, this important publication was originally compiled by the Oregon Department of Geology and Mineral Industries includes detailed descriptions, histories and the geology of over 500 metallic mineral mines and prospects in Josephine County, Oregon. **8.5" X 11", 250 ppgs. Retail Price: $24.99**

Metal Mines of North East Oregon - Oregon mining historian Kerby Jackson introduces us to a classic work on Oregon's mining history in this important re-issue of Bulletin 14A and 14B, otherwise known as the North East Oregon Metal Mines Handbook. Unavailable since 1941, this important publication was originally compiled by the Oregon Department of Geology and Mineral Industries and includes detailed descriptions, histories and the geology of over 750 metallic mineral mines and prospects in North Eastern Oregon. **8.5" X 11", 310 ppgs. Retail Price: $29.99**

Metal Mines of North West Oregon - Oregon mining historian Kerby Jackson introduces us to a classic work on Oregon's mining history in this important re-issue of Bulletin 14D, otherwise known as the North West Oregon Metal Mines Handbook. Unavailable since 1951, this important publication was originally compiled by the Oregon Department of Geology and Mineral Industries and includes detailed descriptions, histories and the geology of over 250 metallic mineral mines and prospects in North Western Oregon. **8.5" X 11", 182 ppgs. Retail Price: $19.99**

Mines and Prospects of Oregon - Mining historian Kerby Jackson introduces us to a classic mining work by the Oregon Bureau of Mines in this important re-issue of The Handbook of Mines and Prospects of Oregon. Unavailable since 1916, this publication includes important insights into hundreds of gold, silver, copper, coal, limestone and other mines that operated in the State of Oregon around the turn of the 19th Century. Included are not only geological details on early mines throughout Oregon, but also insights into their history, production, locations and in some cases, also included are rare maps of their underground workings. **8.5" X 11", 314 ppgs. Retail Price: $24.99**

Lode Gold of the Klamath Mountains of Northern California and South West Oregon
(See California Mining Books)

Mineral Resources of South West Oregon - Unavailable since 1914, this publication includes important insights into dozens of mines that once operated in South West Oregon, including the famous gold fields of Josephine and Jackson Counties, as well as the Coal Mines of Coos County. Included are not only geological details on early mines throughout South West Oregon, but also insights into their history, production and locations. **8.5" X 11", 154 ppgs. Retail Price: $11.99**

Chromite Mining in The Klamath Mountains of California and Oregon
(See California Mining Books)

Southern Oregon Mineral Wealth - Unavailable since 1904, this rare publication provides a unique snapshot into the mines that were operating in the area at the time. Included are not only geological details on early mines throughout South West Oregon, but also insights into their history, production and locations. Some of the mining areas include Grave Creek, Greenback, Wolf Creek, Jump Off Joe Creek, Granite Hill, Galice, Mount Reuben, Gold Hill, Galls Creek, Kane Creek, Sardine Creek, Birdseye Creek, Evans Creek, Foots Creek, Jacksonville, Ashland, the Applegate River, Waldo, Kerby and the Illinois River, Althouse and Sucker Creek, as well as insights into local copper mining and other topics. **8.5" X 11", 64 ppgs. Retail Price: $8.99**

Geology and Ore Deposits of the Takilma and Waldo Mining Districts - Unavailable since the 1933, this publication was originally compiled by the United States Geological Survey and includes details on gold and copper mining in the Takilma and Waldo Districts of Josephine County, Oregon. The Waldo and Takilma mining districts are most notable for the fact that the earliest large scale mining of placer gold and copper in Oregon took place in these two areas. Included in this report are details about some of the earliest large gold mines in the state such as the Llano de Oro, High Gravel, Cameron, Platerica, Deep Gravel and others, as well as copper mines such as the famous Queen of Bronze mine, the Waldo, Lily and Cowboy mines. In addition to geological examinations, insights are also provided into the production, day to day operations and early histories of these mines, as well as calculations of known mineral reserves in the area. This volume also includes six maps and 20 original illustrations. **8.5" X 11", 74 ppgs. Retail Price: $9.99**

Gold Mines of Oregon - Oregon mining historian Kerby Jackson introduces us to a classic work on Oregon's mining history in this important re-issue of Bulletin 61, otherwise known as "Gold and Silver In Oregon". Unavailable since 1968, this important publication was originally compiled by geologists Howard C. Brooks and Len Ramp of the Oregon Department of Geology and Mineral Industries and includes detailed descriptions, histories and the geology of over 450 gold mines Oregon. Included are notes on the history, geology and gold production statistics of all the major mining areas in Oregon including the Klamath Mountains, the Blue Mountains and the North Cascades. While gold is where you find it, as every miner knows, the path to success is to prospect for gold where it was previously found. **8.5″ X 11″, 344 ppgs. Retail Price: $24.99**

Mines and Mineral Resources of Curry County Oregon - Originally published in 1916, this important publication on Oregon Mining has not been available for nearly a century. Included are rare insights into the history, production and locations of dozens of gold mines in Curry County, Oregon, as well as detailed information on important Oregon mining districts in that area such as those at Agness, Bald Face Creek, Mule Creek, Boulder Creek, China Diggings, Collier Creek, Elk River, Gold Beach, Rock Creek, Sixes River and elsewhere. Particular attention is especially paid to the famous beach gold deposits of this portion of the Oregon Coast. **8.5″ X 11″, 140 ppgs. Retail Price: $11.99**

Chromite Mining in South West Oregon - Originally published in 1961, this important publication on Oregon Mining has not been available for nearly a century. Included are rare insights into the history, production and locations of nearly 300 chromite mines in South Western Oregon. **8.5″ X 11″, 184 ppgs. Retail Price: $14.99**

Mineral Resources of Douglas County Oregon - Originally published in 1972, this important publication on Oregon Mining has not been available for nearly forty years. Included are rare insights into the geology, history, production and locations of numerous gold mines and other mining properties in Douglas County, Oregon. **8.5″ X 11″, 124 ppgs. Retail Price: $11.99**

Mineral Resources of Coos County Oregon - Originally published in 1972, this important publication on Oregon Mining has not been available for nearly forty years. Included are rare insights into the geology, history, production and locations of numerous gold mines and other mining properties in Coos County, Oregon. **8.5″ X 11″, 100 ppgs. Retail Price: $11.99**

Mineral Resources of Lane County Oregon - Originally published in 1938, this important publication on Oregon Mining has not been available for nearly seventy five years. Included are extremely rare insights into the geology and mines of Lane County, Oregon, in particular in the Bohemia, Blue River, Oakridge, Black Butte and Winberry Mining Districts. **8.5″ X 11″, 82 ppgs. Retail Price: $9.99**

Mineral Resources of the Upper Chetco River of Oregon: Including the Kalmiopsis Wilderness - Originally published in 1975, this important publication on Oregon Mining has not been available for nearly forty years. Withdrawn under the 1872 Mining Act since 1984, real insight into the minerals resources and mines of the Upper Chetco River has long been unavailable due to the remoteness of the area. Despite this, the decades of battle between property owners and environmental extremists over the last private mining inholding in the area has continued to pique the interest of those interested in mining and other forms of natural resource use. Gold mining began in the area in the 1850's and has a rich history in this geographic area, even if the facts surrounding it are little known. Included are twenty two rare photographs, as well as insights into the Becca and Morning Mine, the Emmly Mine (also known as Emily Camp), the Frazier Mine, the Golden Dream or Higgins Mine, Hustis Mine, Peck Mine and others. **8.5″ X 11″, 64 ppgs. Retail Price: $8.99**

Gold Dredging in Oregon - Originally published in 1939, this important publication on Oregon Mining has not been available for nearly seventy five years. Included are extremely rare insights into the history and day to day operations of the dragline and bucketline gold dredges that once worked the placer gold fields of South West and North East Oregon in decades gone by. Also included are details into the areas that were worked by gold dredges in Josephine, Jackson, Baker and Grant counties, as well as the economic factors that impacted this mining method. This volume also offers a unique look into the values of river bottom land in relation to both farming and mining, in how farm lands were mined, re-soiled and reclamated after the dredges worked them. Featured are hard to find maps of the gold dredge fields, as well as rare photographs from a bygone era. **8.5″ X 11″, 86 ppgs. Retail Price: $8.99**

Quick Silver Mining in Oregon - Originally published in 1963, this important publication on Oregon Mining has not been available for over fifty years. This publication includes details into the history and production of Elemental Mercury or Quicksilver in the State of Oregon. **8.5″ X 11″, 238 ppgs. Retail Price: $15.99**

Mines of the Greenhorn Mining District of Grant County Oregon - Originally published in 1948, this important publication on Oregon Mining has not been available for over sixty five years. In this publication are rare insights into the mines of the famous Greenhorn Mining District of Grant County, Oregon, especially the famous Morning Mine. Also included are details on the Tempest, Tiger, Bi-Metallic, Windsor, Psyche, Big Johnny, Snow Creek, Banzette and Paramount Mines, as well as prospects in the vicinities in the famous mining areas of Mormon Basin, Vinegar Basin and Desolation Creek. Included are hard to find mine maps and dozens of rare photographs from the bygone era of Grant County's rich mining history. **8.5″ X 11″, 72 ppgs. Retail Price: $9.99**

Geology of the Wallowa Mountains of Oregon: Part I (Volume 1) - Originally published in 1938, this important publication on Oregon Mining has not been available for nearly seventy five years. Included are details on the geology of this unique portion of North Eastern Oregon. This is the first part of a two book series on the area. Accompanying the text are rare photographs and historic maps.**8.5" X 11", 92 ppgs. Retail Price: $9.99**

Geology of the Wallowa Mountains of Oregon: Part II (Volume 2) - Originally published in 1938, this important publication on Oregon Mining has not been available for nearly seventy five years. Included are details on the geology of this unique portion of North Eastern Oregon. This is the first part of a two book series on the area. Accompanying the text are rare photographs and historic maps.**8.5" X 11", 94 ppgs. Retail Price: $9.99**

Field Identification of Minerals For Oregon Prospectors - Originally published in 1940, this important publication on Oregon Mining has not been available for nearly seventy five years. Included in this volume is an easy system for testing and identifying a wide range of minerals that might be found by prospectors, geologists and rockhounds in the State of Oregon, as well as in other locales. Topics include how to put together your own field testing kit and how to conduct rudimentary tests in the field. This volume is written in a clear and concise way to make it useful even for beginners. **8.5" X 11", 158 ppgs. Retail Price: $14.99**

The Bohemia Mining District of Oregon - Originally published in 1900, this important publication on Oregon Mining has not been available for over a century. Included in this volume are important insights into the famous Bohemia Mining District of Oregon, including the histories and locations of important gold mines in the area such as the Ophir Mine, Clarence, Acturas, Peek-a-boo, White Swan, Combination Mine, the Musick Mine, The California, White Ghost, The Mystery, Wall Street, Vesuvius, Story, Lizzie Bullock, Delta, Elsie Dora, Golden Slipper, Broadway, Champion Mine, Knott, Noonday, Helena, White Wings, Riverside and others. Also included are notes on the nearby Blue River Mining District. **8.5" X 11", 58 ppgs. Retail Price: $9.99**

The Gold Fields of Eastern Oregon - Unavailable since 1900, this publication was originally compiled by the Baker City Chamber of Commerce Offering important insights into the gold mining history of Eastern Oregon, "The Gold Fields of Eastern Oregon" sheds a rare light on many of the gold mines that were operating at the turn of the 19th Century in Baker County and Grant County in North Eastern Oregon. Some of the areas featured include the Cable Cove District, Baisely-Elhorn, Granite, Red Boy, Bonanza, Susanville, Sparta, Virtue, Vaughn, Sumpter, Burnt River, Rye Valley and other mining districts. Included is basic information on not only many gold mines that are well known to those interested in Eastern Oregon mining history, but also many mines and prospects which have been mostly lost to the passage of time. Accompanying are numerous rare photos **8.5" X 11", 78 ppgs. Retail Price: $10.99**

Gold Mining in Eastern Oregon - Originally published in 1938, this important publication on Oregon Mining has not been available for over a century. Included in this volume are important insights into the famous mining districts of Eastern Oregon during the late 1930's. Particular attention is given to those gold mines with milling and concentrating facilities in the Greenhorn, Red Boy, Alamo, Bonanza, Granite, Cable Cove, Cracker Creek, Virtue, Keating, Medical Springs, Sanger, Sparta, Chicken Creek, Mormon Basin, Connor Creek, Cornucopia and the Bull Run Mining Districts. Some of the mines featured include the Ben Harrison, North Pole-Columbia, Highland Maxwell, Baisley-Elkhorn, White Swan, Balm Creek, Twin Baby, Gem of Sparta, New Deal, Gleason, Gifford-Johnson, Cornucopia, Record, Bull Run, Orion and others. Of particular interest are the mill flow sheets and descriptions of milling operations of these mines. **8.5" X 11", 68 ppgs. Retail Price: $8.99**

The Gold Belt of the Blue Mountains of Oregon - Originally published in 1901, this important publication on Oregon Mining has not been available for over a century. Included in this volume are rare insights into the gold deposits of the Blue Mountains of North East Oregon, including the history of their early discovery and early production. Extensive details are offered on this important mining area's mineralogy and economic geology, as well as insights into nearby gold placers, silver deposits and copper deposits. Featured are the Elkhorn and Rock Creek mining districts, the Pocahontas district, Auburn and Minersville districts, Sumpter and Cracker Creek, Cable Cove, the Camp Carson district, Granite, Alamo, Greenhorn, Robinsonville, the Upper Burnt River Valley and Bonanza districts, Susanville, Quartzburg, Canyon Creek, Virtue, the Copper Butte district, the North Powder River, Sparta, Eagle Creek, Cornucopia, Pine Creek, Lower Powder River, the Upper Snake River Canyon, Rye Valley, Lower Burnt River Valley, Mormon Basin, the Malheur and Clarks Creek districts, Sutton Creek and others. Of particular interest are important details on numerous gold mines and prospects in these mining districts, including their locations, histories, geology and other important information, as well as information on silver, copper and fire opal deposits. **8.5" X 11", 250 ppgs. Retail Price: $24.99**

Mining in the Cascades Range of Oregon - Originally published in 1938, this important publication on Oregon Mining has not been available for over seventy five years. Included in this volume are rare insights into the gold mines and other types of metal mines in the Cascades Mountain Range of Oregon. Some of the important mining areas covered include the famous Bohemia Mining District, the North Santiam Mining District, Quartzville Mining District, Blue River Mining District, Fall Creek Mining District, Oakridge District, Zinc District, Buzzard-Al Sarena District, Grand Cove, Climax District and Barron Mining District. Of particular interest are important details on over 100 mines and prospects in these mining districts, including their locations, histories, geology and other important information. 8.5" X 11", 170 ppgs. Retail Price: $14.99

Beach Gold Placers of the Oregon Coast - Originally published in 1934, this important publication on Oregon Mining has not been available for over 80 years. Included in this volume are rare insights into the beach gold deposits of the State of Oregon, including their locations, occurance, composition and geology. Of particular interest is information on placer platinum in Oregon's rich beach deposits. Also included are the locations and other information on some famous Oregon beach mines, including the Pioneer, Eagle, Chickamin, Iowa and beach placer mines north of the mouth of the Rogue River. 8.5" X 11", 60 ppgs. Retail Price: $8.99

Mineralogical Composition of the Sands of the Oregon Coast: From Coos Bay to the Columbia - Published in 1945, he text features hard to find information on the composition of the gold bearing black sands of the South West Oregon Coast, offering a unique insight to prospectors in search of Oregon's legendary beach gold. 104 ppgs, $9.99

Manganese Mining in Oregon - First released in 1942 and now out of print, this special reprint edition of "Manganese in Oregon" was originally published by the Oregon Department of Geology and Mineral Industries. The text features hard to find information on the mining of Manganese in Oregon, including details and maps of Oregon manganese mines and prospects. 108 ppgs, 9.99

Medford Oregon As A Mining Center - Written in 1912, this hard to find publication includes valuable insights into the mining history of South West Oregon. This small book contains interesting information on the gold, copper and mining industry in Southern Oregon as it existed just prior to World War One, shedding light on some of the important mines in the area. Included are rare photographs and vintage advertising of the day. 80 ppgs, 9.99

Mineral Resources of Curry County Oregon - First released in 1977 and now out of print, this special reprint edition of "Geology, Mineral Resources and Rock Materials of Curry County, Oregon" was originally published in cooperation of Curry County, Oregon and the Oregon Department of Geology and Mineral Industries. The text features hard to find information on not only the mining of gold and other metals in Curry County, but also aggregate mining in the area. 102 ppgs, 11.99

Origin of the Gold Bearing Black Sands of the Coast of South West Oregon - First released in 1943 and now out of print, this special reprint edition of "The Origin of the Black Sands of the South West Oregon Coast" was originally published by the Oregon Department of Geology and Mineral Industries. The text features hard to find information on the origin of the gold bearing black sands of the South West Oregon Coast, offering a unique insight to prospectors in search of Oregon's legendary beach gold. 52 ppgs, 8.99

South West Oregon Mining - Leading mining historian Kerby Jackson introduces us to six classic small mining publications on the Gold Mining Industry in Southern Oregon. This small book consists of a compilation of USGS J.S. Diller's "Mines of the Riddles Quadrangle", "The Rogue River Valley Coal Fields" and "Mineral Resources of the Grants Pass Quadrangle", the Grants Pass Commercial Club's rare publication "Mining in Josephine County, Oregon" and the USGS publication "The Distribution of Placer Gold in the Sixes River, South West Oregon". Also included is F.W. Libbey's legendary article on the Southern Oregon Mining Industry, "Lest We Forget", which appeared in the publication of the Oregon State Department of Geology and Mineral Industries in the early 1960's. This compilation offers a unique perspective on mining in South West Oregon and includes considerable information on mines in Josephine, Jackson and Coos Counties. 142 ppgs, 14.99

Geology and Mineral Resources of the Gasquet Quadrangle of California-Oregon - First published in 1953, it has been unavailable for over a century and sheds important light on the geological features and mineral resources of this portion of Northern California and Southern Oregon. 80 ppgs, 9.99

Idaho Mining Books

Gold in Idaho - Unavailable since the 1940's, this publication was originally compiled by the Idaho Bureau of Mines and includes details on gold mining in Idaho. Included is not only raw data on gold production in Idaho, but also valuable insight into where gold may be found in Idaho, as well as practical information on the gold bearing rocks and other geological features that will assist those looking for placer and lode gold in the State of Idaho. This volume also includes thirteen gold maps that greatly enhance the practical usability of the information contained in this small book detailing where to find gold in Idaho. **8.5" X 11", 72 ppgs. Retail Price: $9.99**

Geology of the Couer D'Alene Mining District of Idaho - Unavailable since 1961, this publication was originally compiled by the Idaho Bureau of Mines and Geology and includes details on the mining of gold, silver and other minerals in the famous Coeur D'Alene Mining District in Northern Idaho. Included are details on the early history of the Coeur D'Alene Mining District, local tectonic settings, ore deposit features, information on the mineral belts of the Osburn Fault, as well as detailed information on the famous Bunker Hill Mine, the Dayrock Mine, Galena Mine, Lucky Friday Mine and the infamous Sunshine Mine. This volume also includes sixteen hard to find maps. **8.5" X 11", 70 ppgs. Retail Price: $9.99**

The Gold Camps and Silver Cities of Idaho - Originally published in 1963, this important publication on Idaho Mining has not been available for nearly fifty years. Included are rare insights into the history of Idaho's Gold Rush, as well as the mad craze for silver in the Idaho Panhandle. Documented in fine detail are the early mining excitements at Boise Basin, at South Boise, in the Owyhees, at Deadwood, Long Valley, Stanley Basin and Robinson Bar, at Atlanta, on the famous Boise River, Volcano, Little Smokey, Banner, Boise Ridge, Hailey, Leesburg, Lemhi, Pearl, at South Mountain, Shoup and Ulysses, Yellow Jacket and Loon Creek. The story follows with the appearance of Chinese miners at the new mining camps on the Snake River, Black Pine, Yankee Fork, Bay Horse, Clayton, Heath, Seven Devils, Gibbonsville, Vienna and Sawtooth City. Also included are special sections on the Idaho Lead and Silver mines of the late 1800's, as well as the mining discoveries of the early 1900's that paved the way for Idaho's modern mining and mineral industry. Lavishly illustrated with rare historic photos, this volume provides a one of a kind documentary into Idaho's mining history that is sure to be enjoyed by not only modern miners and prospectors who still scour the hills in search of nature's treasures, but also those enjoy history and tromping through overgrown ghost towns and long abandoned mining camps. **8.5" X 11", 186 ppgs. Retail Price: $14.99**

Ore Deposits and Mining in North Western Custer County Idaho - Unavailable since 1913, this important publication was originally published by the Us Department of the Interior and has been unavailable for a century. Included are fine details on the geology, geography, gold placers and gold and silver bearing quartz veins of the mining region of North West Custer County, Idaho. Of particular interest is a rare look at the mines and prospects of the region, including those such as the Ramshorn Mine, SkyLark, Riverview, Excelsior, Beardsley, Pacific, Hoosier, Silver Brick, Forest Rose and dozens of others in the Bay Horse Mining District. Also covered are the mines of the Yankee Fork District such as the Lucky Boy, Badger, Black, Enterprise, Charles Dickens, Morrison, Golden Sunbeam, Montana, Golden Gate and others, as well as those in the Loon Mining District. **8.5" X 11", 126 ppgs. Retail Price: $12.99**

Gold Rush To Idaho - Unavailable since 1963, this important publication was originally published by the Idaho Bureau of Mines and has been unavailable for 50 years. "Gold Rush To Idaho" revisits the earliest years of the discovery of gold in Idaho Territory and introduces us to the conditions that the pioneer gold seekers met when they blazed a trail through the wilderness of Idaho's mountains and discovered the precious yellow metal at Oro Fino and Pierce. Subsequent rushes followed at places like Elk City, Newsome, Clearwater Station, Florence, Warrens and elsewhere. Of particular interest is a rare look at the hardships that the first miners in Idaho met with during their day to day existences and their attempts to bring law and order to their mining camps. **8.5" X 11", 88 ppgs. Retail Price: $9.99**

The Geology and Mines of Northern Idaho and North Western Montana - Unavailable since 1909, this important publication was originally published by the Us Department of the Interior and has been unavailable for a century. Included are fine details on the geology and geography of the mining regions of Northern Idaho and North Western Montana. Of particular interest is a rare look at the mines and prospects of the region, including those in the Pine Creek Mining District, Lake Pend Oreille district, Troy Mining District, Sylvanite District, Cabinet Mining District, Prospect Mining District and the Missoula Valley. Some of the mines featured include the Iron Mountain, Silver Butte, Snowshoe, Grouse Mountain Mine and others. **8.5" X 11", 142 ppgs. Retail Price: $12.99**

Mining in the Alturas Quadrangle of Blaine County Idaho - Unavailable since 1922, this important publication was originally published by the Idaho Bureau of Mines and has been unavailable for ninety years. Topics include the geology, rock formations and the formation of ore deposits in this important mining area of Idaho. Of particular focus is information on the local geology, quartz veins and ore deposits of this portion of Idaho. Included are hard to find details, including the descriptions and locations of numerous gold and silver mines in the area including the Silver King, Pilgrim, Columbia, Lone Jack, Sunbeam, Pride of the West, Lucky Boy, Scotia, Atlanta, Beaver-Bidwell and others mines and prospects. **8.5" X 11", 56 ppgs. Retail Price: $8.99**

Mining in Lemhi County Idaho - Originally published in 1913, this important book on Idaho Mining has not been available to miners for over a century. Included are rare insights into hundreds of gold, silver, copper and other mines in this famous Idaho mining area. Details include the locations, geology, history, production and other facts of the mines of this region, not only gold and silver hardrock mines, but also gold placer mines, lead-silver deposits, copper mines, cobalt-nickel deposits, tungsten and tin mines . It is lavishly illustrated with hard to find photos of the period and rare mining maps. Some of the vicinities featured include the Nicholia Mining District, Spring Mountain District, Texas District, Blue Wing District, Junction District, McDevitt District, Pratt Creek, Eldorado District, Kirtley Creek, Carmen Creek, Gibbonsville, Indian Creek, Mineral Hill District, Mackinaw, Eureka District, Blackbird District, YellowJacket District, Gravel Range District, Junction District, Parker Mountain and other mining districts. **8.5" X 11", 226 ppgs. Retail Price: $19.99**

Mining in Shoshone County Idaho - First published in 1923, it has been unavailable for over a century and sheds important light on the mining history of Shoshone County, Idaho. Some of the topics include the history of mining in Shoshone County, a look at the local geology and ore characteristics of lead-silver deposits, zinc deposits, copper, antimony, gold and other minerals. Also included are insights into the history, production, characteristics and locations of numerous mines in the area. 198 ppgs, 15.99

Utah Mining Books

Fluorite in Utah - Unavailable since 1954, this publication was originally compiled by the USGS, State of Utah and U.S. Atomic Energy Commission and details the mining of fluorspar, also known as fluorite in the State of Utah. Included are details on the geology and history of fluorspar (fluorite) mining in Utah, including details on where this unique gem mineral may be found in the State of Utah. **8.5" X 11", 60 ppgs. Retail Price: $8.99**

The Gold Hill Mining District of Utah - First published in 1935, it has been unavailable since those days and sheds important light on the mines, history and geology of Utah's Gold Hill Mining District. Included are rare insights into this important mining area, including the locations, histories and details of numerous mines. This volume is well illustrated with geological diagrams, as well as hard to find maps of some of the most important mines in this district. 202 ppgs., 19.99

The Mines, Miners and Minerals of Utah - First published in 1896, it has been unavailable since those days and sheds important light on the early mines and miners of Pioneer Utah, as well as the minerals which they won from the earth by laborious hard physical labor and sheer determination. Included are rare insights into the early mining history of Utah, as well details on hundreds of gold, silver and copper mines. 376 ppgs., 24.99

California Mining Books

The Tertiary Gravels of the Sierra Nevada of California - Mining historian Kerby Jackson introduces us to a classic mining work by Waldemar Lindgren in this important re-issue of The Tertiary Gravels of the Sierra Nevada of California. Unavailable since 1911, this publication includes details on the gold bearing ancient river channels of the famous Sierra Nevada region of California. **8.5" X 11", 282 ppgs. Retail Price: $19.99**

The Mother Lode Mining Region of California - Unavailable since 1900, this publication includes details on the gold mines of California's famous Mother Lode gold mining area. Included are details on the geology, history and important gold mines of the region, as well as insights into historic mining methods, mine timbering, mining machinery, mining bell signals and other details on how these mines operated. Also included are insights into the gold mines of the California Mother Lode that were in operation during the first sixty years of California's mining history. **8.5" X 11", 176 ppgs. Retail Price: $14.99**

Lode Gold of the Klamath Mountains of Northern California and South West Oregon - Unavailable since 1971, this publication was originally compiled by Preston E. Hotz and includes details on the lode mining districts of Oregon and California's Klamath Mountains. Included are details on the geology, history and important lode mines of the French Gulch, Deadwood, Whiskeytown, Shasta, Redding, Muletown, South Fork, Old Diggings, Dog Creek (Delta), Bully Choop (Indian Creek), Harrison Gulch, Hayfork, Minersville, Trinity Center, Canyon Creek, East Fork, New River, Denny, Liberty (Black Bear), Cecilville, Callahan, Yreka, Fort Jones and Happy Camp mining districts in California, as well as the Ashland, Rogue River, Applegate, Illinois River, Takilma, Greenback, Galice, Silver Peak, Myrtle Creek and Mule Creek districts of South Western Oregon. Also included are insights into the mineralization and other characteristics of this important mining region. **8.5" X 11", 100 ppgs. Retail Price: $10.99**

Mines and Mineral Resources of Shasta County, Siskiyou County, Trinity County: California - Unavailable since 1915, this publication was originally compiled by the California State Mining Bureau and includes details on the gold mines of this area of Northern California. Also included are insights into the mineralization and other characteristics of this important mining region, as well as the location of historic gold mines. **8.5" X 11", 204 ppgs. Retail Price: $19.99**

Geology of the Yreka Quadrangle, Siskiyou County, California - Unavailable since 1977, this publication was originally compiled by Preston E. Hotz and includes details on the geology of the Yreka Quadrangle of Siskiyou County, California. Also included are insights into the mineralization and other characteristics of this important mining region. **8.5" X 11", 78 ppgs. Retail Price: $7.99**

Mines of San Diego and Imperial Counties, California - Originally published in 1914, this important publication on California Mining has not been available for a century. This publication includes important information on the early gold mines of San Diego and Imperial County, which were some of the first gold fields mined in California by early Spanish and Mexican miners before the 49ers came on the scene. Included are not only details on early mining methods in the area, production statistics and geological information, but also the location of the early gold mines that helped make California "The Golden State". Also included are details on the mining of other minerals such as silver, lead, zinc, manganese, tungsten, vanadium, asbestos, barite, borax, cement, clay, dolomite, fluospar, gem stones, graphite, marble, salines, petroleum, stronium, talc and others. **8.5" X 11", 116 ppgs. Retail Price: $12.99**

Mines of Sierra County, California - Unavailable since 1920, this publication was originally compiled by the California State Mining Bureau and includes details on the gold mines of Sierra County, California. Also included are insights into the mineralization and other characteristics of this important mining region, as well as the location of historic gold mines. **8.5" X 11", 156 ppgs. Retail Price: $19.99**

Mines of Plumas County, California - Unavailable since 1918, this publication was originally compiled by the California State Mining Bureau and includes details on the gold mines of Plumas County, California. Also included are insights into the mineralization and other characteristics of this important mining region, as well as the location of historic gold mines. **8.5" X 11", 200 ppgs. Retail Price: $19.99**

Mines of El Dorado, Placer, Sacramento and Yuba Counties, California - Originally published in 1917, this important publication on California Mining has not been available for nearly a century. This publication includes important information on the early gold mines of El Dorado County, Placer County, Sacramento County and Yuba County, which were some of the first gold fields mined by the Forty-Niners during the California Gold Rush. Included are not only details on early mining methods in the area, production statistics and geological information, but also the location of the early gold mines that helped make California "The Golden State". Also included are insights into the early mining of chrome, copper and other minerals in this important mining area. **8.5" X 11", 204 ppgs. Retail Price: $19.99**

Mines of Los Angeles, Orange and Riverside Counties, California - Originally published in 1917, this important publication on California Mining has not been available for nearly a century. This publication includes important information on the early gold mines of Los Angeles County, Orange County and Riverside County, which were some of the first gold fields mined in California by early Spanish and Mexican miners before the 49ers came on the scene. Included are not only details on early mining methods in the area, production statistics and geological information, but also the location of the early gold mines that helped make California "The Golden State". **8.5" X 11", 146 ppgs. Retail Price: $12.99**

Mines of San Bernadino and Tulare Counties, California - Originally published in 1917, this important publication on California Mining has not been available for nearly a century. This publication includes important information on the early gold mines of San Bernadino and Tulare County, which were some of the first gold fields mined in California by early Spanish and Mexican miners before the 49ers came on the scene. Included are not only details on early mining methods in the area, production statistics and geological information, but also the location of the early gold mines that helped make California "The Golden State". Also included are details on the mining of other minerals such as copper, iron, lead, zinc, manganese, tungsten, vanadium, asbestos, barite, borax, cement, clay, dolomite, fluospar, gem stones, graphite, marble, salines, petroleum, stronium, talc and others. **8.5" X 11", 200 ppgs. Retail Price: $19.99**

Chromite Mining in The Klamath Mountains of California and Oregon - Unavailable since 1919, this publication was originally compiled by J.S. Diller of the United States Department of Geological Survey and includes details on the chromite mines of this area of Northern California and Southern Oregon. Also included are insights into the mineralization and other characteristics of this important mining region, as well as the location of historic mines. Also included are insights into chromite mining in Eastern Oregon and Montana. **8.5" X 11", 98 ppgs. Retail Price: $9.99**

Mines and Mining in Amador, Calaveras and Tuolumne Counties, California - Unavailable since 1915, this publication was originally compiled by William Tucker and includes details on the mines and mineral resources of this important California mining area. Included are details on the geology, history and important gold mines of the region, as well as insights into other local mineral resources such as asbestos, clay, copper, talc, limestone and others. Also included are insights into the mineralization and other characteristics of this important portion of California's Mother Lode mining region. **8.5" X 11", 198 ppgs. Retail Price: $14.99**

The Cerro Gordo Mining District of Inyo County California - Unavailable since 1963, this publication was originally compiled by the United States Department of Interior. Included are insights into the mineralization and other characteristics of this important mining region of Southern California. Topics include the mining of gold and silver in this important mining district in Inyo County, California, including details on the history, production and locations of the Cerro Gordo Mine, the Morning Star Mine, Estelle Tunnel, Charles Lease Tunnel, Ignacio, Hart, Crosscut Tunnel, Sunset, Upper Newtown, Newtown, Ella, Perseverance, Newsboy, Belmont and other silver and gold mines in the Cerro Gordo Mining District. This volume also includes important insights into the fossil record, geologic formations, faults and other aspects of economic geology in this California mining district. 8.5" X 11", 104 ppgs. Retail Price: $10.99

Mining in Butte, Lassen, Modoc, Sutter and Tehama Counties of California - Unavailable since 1917, this publication was originally compiled by the United States Department of Interior. Included are insights into the mineralization and other characteristics of this important mining region of California. Topics include the mining of asbestos, chromite, gold, diamonds and manganese in Butte County, the mining of gold and copper in the Hayden Hill and Diamond Mountain mining districts of Lassen County, the mining of coal, salt, copper and gold in the High Grade and Winters mining districts of Modoc County, gold mining in Sutter County and the mining of gold, chromite, manganese and copper in Tehama County. This volume also includes the production records and locations of numerous mines in this important mining region. 8.5" X 11", 114 ppgs. Retail Price: $11.99

Mines of Trinity County California - Originally published in 1965, this important publication on California Mining has not been available for nearly fifty years. This publication includes important information on mines and mining in Trinity County, California, as well insights into the mineralization and geology of this important mining area in Northern California. Included are extensive details on hardrock and placer gold mines and prospects, including charts showing the locations of these historic mines.. 8.5" X 11", 144 ppgs. Retail Price: $12.99

Mines of Kern County California - Originally published in 1962, this important publication on California Mining has not been available for nearly fifty years. This publication includes important information on mines and mining in Kern County, California, as well insights into the mineralization and geology of this important mining area in California. Included are extensive details on hardrock and placer gold mines and prospects, including charts showing the locations of these historic mines. 8.5" X 11", 398 ppgs. Retail Price: $24.99

Mines of Calaveras County California - Originally published in 1962, this important publication on California Mining has not been available for nearly fifty years. This publication includes important information on mines and mining in Calaveras County, California, as well insights into the mineralization and geology of this important mining area in Northern California. Included are extensive details on hardrock and placer gold mines and prospects, including charts showing the locations of these historic mines. 8.5" X 11", 236 ppgs. Retail Price: $19.99

Lode Gold Mining in Grass Valley California - Unavailable since 1940, this publication was originally compiled by the United States Department of Interior. Included are insights into the gold mineralization and other characteristics of this important mining region of Nevada County, California. This volume also includes important insights into the geologic formations, faults and other aspects of economic geology in this California mining district. Of particular interest are the fine details on many hardrock gold mines in the area, including their locations, histories, development and mineralization. Some of the mines featured include the Gold Hill Mine, Massachusetts Hill, Boundary, Peabody, Golden Center, North Star, Omaha, Lone Jack, Homeward Bound, Hartery, Wisconsin, Allison Ranch, Phoenix, Kate Hayes, W.Y.O.D., Empire, Rich Hill, Daisy Hill, Orleans, Sultana, Centennial, Conlin, Ben Franklin, Crown Point and many others. 8.5" X 11", 148 ppgs. Retail Price: $12.99

Lode Mining in the Alleghany District of Sierra County California - Unavailable since 1913, this publication was originally compiled by the United States Department of Interior. Included are insights into the mineralization and other characteristics of this important mining region of Sierra County. Included are details on the history, production and locations of numerous hardrock gold mines in this famous California area, including the Tightner Mine, Minnie D., Osceola, Eldorado, Twenty One, Sherman, Kenton, Oriental, Rainbow, Plumbago, Irelan, Gold Canyon, North Fork, Federal, Kate Hardy and others. This volume also includes important insights into the fossil record, geologic formations, faults and other aspects of economic geology in this California mining district. 8.5" X 11", 48 ppgs. Retail Price: $7.99

Six Months In The Gold Mines During The California Gold Rush - Unavailable since 1850, this important work is a first hand account of one "49'ers" personal experience during the great California Gold Rush, shedding important light on one of the most exciting periods in the history of not only California, but also the world. Compiled from journals written between 1847 and 1849 by E. Gould Buffum, a native of New York, "Six Months In The Gold Mines During The California Gold Rush" offers a rare look into the day to day lives of the people who came to California to work in her gold mines when the state was still a great frontier. 8.5" X 11", 290 ppgs. Retail Price: $19.99

Quartz Mines of the Grass Valley Mining District of California - Unavailable since 1867, this important publication has not been available since those days. This rare publication offers a short dissertation on the early hardrock mines in this important mining district in the California Mother Lode region between the 1850's and 1860's. Also included are hard to find details on the mineralization and locations of these mines, as well as how they were operated in those day. **8.5" X 11", 44 ppgs. Retail Price: $8.99**

Gold Rush on the Feather River - First published in 1924, this short publication by G.C. Mansfield sheds important light on the early history of gold mining on the Feather River. Included are rare insights into the first decade of gold mining and the early mining camps of the Feather River during the 1850's. 64 ppgs., 9.99

The Bodie Mining District of California - First published in 1986, it has been unavailable since those days and sheds important light on this famous mining area. Included are the history, characteristics and locations of numerous old mines around the ghost town of Bodie. 64 ppgs, 8.99

Geology and Mineral Resources of the Gasquet Quadrangle of California-Oregon - First published in 1953, it has been unavailable for over a century and sheds important light on the geological features and mineral resources of this portion of Northern California and Southern Oregon. 80 ppgs, 9.99

Alaska Mining Books

Ore Deposits of the Willow Creek Mining District, Alaska - Unavailable since 1954, this hard to find publication includes valuable insights into the Willow Creek Mining District near Hatcher Pass in Alaska. The publication includes insights into the history, geology and locations of the well known mines in the area, including the Gold Cord, Independence, Fern, Mabel, Lonesome, Snowbird, Schroff-O'Neil, High Grade, Marion Twin, Thorpe, Webfoot, Kelly-Willow, Lane, Holland and others. **8.5" X 11", 96 ppgs. Retail Price: $9.99**

The Juneau Gold Belt of Alaska - Unavailable since 1906, this hard to find publication includes valuable insights into the gold mines around Juneau, Alaska. The publication includes important details into the history, geology and locations of the well known gold mines and prospects in the area, including those around Windham Bay, Holkham Bay, Port Snettisham, on Grindstone and Rhine Creeks, Gold Creek, Douglas Island, Salmon Creek, Lemon Creek, Nugget Creek, from the Mendenhall River to Berners Bay, McGinnis Creek, Montana Creek, Peterson Creek, Windfall Creek, the Eagle River, Yankee Basin, Yankee Curve, Kowee Creek and elsewhere. Not only are gold placer mines included, but also hardrock gold mines. **8.5" X 11", 224 ppgs. Retail Price: $19.99**

Mining in the Jumbo Basin of Alaska - Unavailable since 1953, this hard to find publication includes valuable insights into the mines and geology of the Jumbo Basin. The publication includes important details into the history, geology and locations of the well known gold mines and prospects in the famous Jumbo Basin Mining Region of Alaska. 72 ppgs, 9.99

The Rampart Placer Gold Region of Alaska - Unavailable since 1906, this hard to find publication includes valuable insights into the placer gold mines of the Rampart Mining Region. The publication includes important details into the history, geology and locations of the well known gold mines and prospects in the famous Rampart Mining Region of Alaska. 78 ppgs, 10.99

Arizona Mining Books

Mines and Mining in Northern Yuma County Arizona - Originally published in 1911, this important publication on Arizona Mining has not been available for over a hundred years. Included are rare insights into the gold, silver, copper and quicksilver mines of Yuma County, Arizona together with hard to find maps and photographs. Some of the mines and mining districts featured include the Planet Copper Mine, Mineral Hill, the Clara Consolidated Mine, Viati Mine, Copper Basin prospect, Bowman Mine, Quartz King, Billy Mack, Carnation, the Wardwell and Osbourne, Valensuella Copper, the Mariquita, Colonial Mine, the French American, the New York-Plomosa, Guadalupe, Lead Camp, Mudersbach Copper Camp, Yellow Bird, the Arizona Northern (Salome Strike), Bonanza (Harqua Hala), Golden Eagle, Hercules, Socorro and others. **8.5" X 11", 144 ppgs. Retail Price: $11.99**

The Aravaipa and Stanley Mining Districts of Graham County Arizona - Originally published in 1925, this important publication on Arizona Mining has not been available for nearly ninety years. Included are rare insights into the gold and silver mines of these two important mining districts, together with hard to find maps. **8.5" X 11", 140 ppgs. Retail Price: $11.99**

Gold in the Gold Basin and Lost Basin Mining Districts of Mohave County, Arizona - This volume contains rare insights into the geology and gold mineralization of the Gold Basin and Lost Basin Mining Districts of Mohave County, Arizona that will be of benefit to miners and prospectors. Also included is a significant body of information on the gold mines and prospects of this portion of Arizona. This volume is lavishly illustrated with rare photos and mining maps. **8.5" X 11", 188 ppgs. Retail Price: $19.99**

Mines of the Jerome and Bradshaw Mountains of Arizona - This important publication on Arizona Mining has not been available for ninety years. This volume contains rare insights into the geology and ore deposits of the Jerome and Bradshaw Mountains of Arizona that will be of benefit to miners and prospectors who work those areas. Included is a significant body of information on the mines and prospects of the Verde, Black Hills, Cherry Creek, Prescott, Walker, Groom Creek, Hassayampa, Bigbug, Turkey Creek, Agua Fria, Black Canyon, Peck, Tiger, Pine Grove, Bradshaw, Tintop, Humbug and Castle Creek Mining Districts. This volume is lavishly illustrated with rare photos and mining maps. **8.5" X 11", 218 ppgs. Retail Price: $19.99**

The Ajo Mining District of Pima County Arizona - This important publication on Arizona Mining has not been available for nearly seventy years. This volume contains rare insights into the geology and mineralization of the Ajo Mining District in Pima County, Arizona and in particular the famous New Cornelia Mine. **8.5" X 11", 126 ppgs. Retail Price: $11.99**

Mining in the Santa Rita and Patagonia Mountains of Arizona - Originally published in 1915, this important publication on Arizona Mining has not been available for nearly a century. Included are rare insights into hundreds of gold, silver, copper and other mines in this famous Arizona mining area. Details include the locations, geology, history, production and other facts of the mines of this region. **8.5" X 11", 394 ppgs. Retail Price: $24.99**

Mining in the Bisbee Quadrangle of Arizona - Originally published in 1906, this important publication on Arizona Mining has not been available for nearly a century. Included are rare insights into hundreds of gold, silver, copper and other mines in this famous Arizona mining area. Details include the locations, geology, history, production and other facts of the mines of this important mining region. **8.5" X 11", 188 ppgs. Retail Price: $14.99**

Placer Gold Mining in Arizona - Unavailable since 1922, this hard to find publication includes valuable insights into the placer gold mines of the Arizona. Originally released as "Placer Gold of Arizona", despite its small size, this publication includes important details into the history, geology and locations of the well known placer gold mines and prospects in the State of Arizona. **48 ppgs, 8.99**

Gold and Copper Mining near Payson, Arizona - Written in 1915, this hard to find publication includes valuable insights into the gold and copper mining industry of Arizona. Highlighted here are the gold and copper mines near Payson, Arizona. **68 ppgs, 8.99**

Lode Gold Mining in Arizona - Unavailable since 1934, this hard to find publication, originally released as "Arizona Lode Gold Mines and Gold Mining" includes valuable insights into the gold mining industry of Arizona. Included are valuable insights into over 150 hardrock gold mines in over 30 different mining districts in Arizona. **278 ppgs, 21.99**

Mining in the Dragoon Quadrangle of Cochise County, Arizona - Unavailable since 1964, this hard to find publication includes valuable insights into the mines of the Dragoon Quadrangle Mining Region. The publication includes important details into the history, geology and locations of the well known mines and prospects in this famous mining region of Arizona. **224 ppgs., 19.99**

Directory of Operating Mines in Arizona in 1915 - Unavailable since 1916, this hard to find publication includes valuable insights into the mines of Arizona. This small publication includes a complete list of the mines that were operating in the State of Arizona during 1915 and includes details such as general location, owners and some basic facts about each mining operation. **52 ppgs. 8.99**

Arizona Ore Deposits - Unavailable since 1938, this hard to find publication includes valuable insights into some ore deposits of Arizona. Included are valuable insights into the formation and characteristics of valuable ore deposits in the Jerome, Miami, Inspiration, Clifton, Morenci, Ray, Ajo, Eureka, Tombstone and Magma mining districts. Included are details into some of the major gold, silver and copper mines of these important Arizona mining areas. **160 ppgs, 14.99**

Montana Mining Books

A History of Butte Montana: The World's Greatest Mining Camp - First published in 1900 by H.C. Freeman, this important publication sheds a bright light on one of the most important mining areas in the history of The West. Together with his insights, as well as rare photographs of the periods, Harry Freeman describes Butte and its vicinity from its early beginnings, right up to its flush years when copper flowed from its mines like a river. At the time of publication, Butte, Montana was known worldwide as "The Richest Mining Spot On Earth" and produced not only vast amounts of copper, but also silver, gold and other metals from its mines. Freeman illustrates, with great detail, the most important mines in the vicinity of Butte, providing rare details on their owners, their history and most importantly, how the mines operated and how their treasures were extracted. Of particular interest are the dozens of rare photographs that depict mines such as the famous Anaconda, the Silver Bow, the Smoke House, Moose, Paulin, Buffalo, Little Minah, the Mountain Consolidated, West Greyrock, Cora, the Green Mountain, Diamond, Bell, Parnell, the Neversweat, Nipper, Original and many others. 8.5" X 11", 142 ppgs. Retail Price: $12.99

The Butte Mining District of Montana - This important publication on Montana Mining has not been available for over a century. Included are rare insights into the gold, copper and silver mines of Butte, Montana together with hard to find maps and photographs. Some of the topics include the early history of gold, silver and copper mining in the Butte area, insight into the geology of its mining areas, the local distribution of gold, silver and copper ores, as well their composition and how to identify them. Also included are detailed facts about the mines in the Butte Mining District, including the famous Anaconda Mine, Gagnon, Parrot, Blue Vein, Moscow, Poulin, Stella, Buffalo, Green Mountain, Wake Up Jim, the Diamond-Bell Group, Mountain Consolidated, East Greyrock, West Greyrock, Snowball, Corra, Speculator, Adirondack, Miners Union, the Jessie-Edith May Group, Otisco, Iduna, Colorado, Lizzie, Cambers, Anderson, Hesperus, Preferencia and dozens of others. 8.5" X 11", 298 ppgs. Retail Price: $24.99

Mines of the Helena Mining Region of Montana - This important publication on Montana Mining has not been available for over a century. Included are rare insights into the gold, copper and silver mines of the vicinity of Helena, Montana, including the Marysville Mining District, Elliston Mining District, Rimini Mining District, Helena Mining District, Clancy Mining District, Wickes Mining District, Boulder and Basin Mining Districts and the Elkhorn Mining District. Some of the topics include the early history of gold, silver and copper mining in the Helena area, insight into the geology of its mining areas, the local distribution of gold, silver and copper ores, as well their composition and how to identify them. Also included are detailed facts, history, geology and locations of over one hundred gold, silver and copper mines in the area . 8.5" X 11", 162 ppgs, Retail Price: $14.99

Mines and Geology of the Garnet Range of Montana - This important publication on Montana Mining has not been available for over a century. Included are rare insights into the gold, copper and silver mines of the vicinity of this important mining area of Montana. Some of the topics include the early history of gold, silver and copper mining in the Garnet Mountains, insight into the geology of its mining areas, the local distribution of gold, silver and copper ores, as well their composition and how to identify them. Also included are detailed facts, history, geology and locations of numerous gold, silver and copper mines in the area . 8.5" X 11", 100 ppgs, Retail Price: $11.99

Mines and Geology of the Philipsburg Quadrangle of Montana - This important publication on Montana Mining has not been available for over a century. Included are rare insights into the gold, copper and silver mines of the vicinity of this important mining area of Montana. Some of the topics include the early history of gold, silver and copper mining in the Philipsburg Quadrangle, insight into the geology of its mining areas, the local distribution of gold, silver and copper ores, as well their composition and how to identify them. Also included are detailed facts, history, geology and locations of over one hundred gold, silver and copper mines in the area 8.5" X 11", 290 ppgs, Retail Price: $24.99

Geology of the Marysville Mining District of Montana - Included are rare insights into the mining geology of the Marysville Mining District. Some of the topics include the early history of gold, silver and copper mining in the area, insight into the geology of its mining areas, the local distribution of gold, silver and copper ores, as well their composition and how to identify them. Also included are detailed facts, history, geology and locations of gold, silver and copper mines in the area 8.5" X 11", 198 ppgs, Retail Price: $19.99

The Geology and Mines of Northern Idaho and North Western Montana- See listing under Idaho.

The History of Gold Dredging in Montana - Unavailable since 1916, this important publication was originally published by the Us Bureau of Mines and has been unavailable for a century. A century and more ago, giant dredging machines dug in Montana's rivers and creeks in search of illusive golden riches. First appearing in California in the 1850's, gold dredges finally reached their peak of development in Siberia and New Zealand before becoming popular again in the United States. This book offers a unique historical perspective on the gold dredges that once operated in Montana. This book on Montana mining history is lavishly illustrated with dozens of rare historic photos gold dredges that once operated in Montana, as well as hard to locate plans on how these dredges were designed. 120 ppgs., 11.99

Nevada Mining Books

The Bull Frog Mining District of Nevada - Unavailable since 1910, this publication was originally compiled by the United States Department of Interior. This volume also includes important insights into the geologic formations, faults and other aspects of economic geology in this Nevada mining district. Of particular interest are the fine details on many mines in the area, including their locations, histories, development and mineralization. Some of the mines featured include the National Bank Mine, Providence, Gibraltor, Tramps, Denver, Original Bullfrog, Gold Bar, Mayflower, Homestake-King and other mines and prospects. **8.5" X 11", 152 ppgs, Retail Price: $14.99**

History of the Comstock Lode - Unavailable since 1876, this publication was originally released by John Wiley & Sons. This volume also includes important insights into the famous Comstock Lode of Nevada that represented the first major silver discovery in the United States. During its spectacular run, the Comstock produced over 192 million ounces of silver and 8.2 million ounces of gold. Not only did the Comstock result in one of the largest mining rushes in history and yield immense fortunes for its owners, but it made important contributions to the development of the State of Nevada, as well as neighboring California. Included here are important details on not only the early development and history of the Comstock, but also rare early insight into its mines, ore and its geology.8.5" X 11", 244 ppgs, Retail Price: $19.99

The Pioche Mining District of Nevada - First published in 1932, it has been unavailable for over a century and sheds important light on the mining history of Nevada. Some of the topics include the history of mining in this district, as well as the characteristics of its mineral and ore deposits. Also included are insights into the history, production, characteristics and locations of numerous mines in the area. Some of the mines include the Combined Metals, Pioche, Ely Valley, No. 10, Poorman, Wide Awake, Alps, Prince, Virginia Louise, Half Moon, Abe Lincoln, Fairview, Bristol Silver, National, Vesuvius, Inman, Tempest, Hillside, Jackrabbit, Lucky Star, Fortuna, Mendha, Manhattan, Hamburg, Comet, Lyndon and others. 108 ppgs 10.99

The Yerington Mining District of Nevada - First published in 1932, it has been unavailable for over a century and sheds important light on the mining history of Nevada. Some of the topics include the history of mining in this district, as well as the characteristics of its mineral and ore deposits. Also included are insights into the history, production, characteristics and locations of numerous mines in the area. Some of the mines include the Bluestone, Mason Valley, Malachite, McConnell, Greenwood, Western Nevada, Ludwig, Douglas Hill, Casting Copper, Montana-Yerington, Empire, Jim Beatty, Terry and McFarland, Blue Jay and others. 92 ppgs, 10.99

The Genesis of the Ores of Tonopah Nevada - Unavailable since 1918, this hard to find publication includes valuable insights into the gold mines around Tonopah, Nevada. The publication includes important details into the geology of mines in the Tonopah Mining District of Nevada. 90 ppgs, 10.99

Mining Camps of Elko, Lander and Eureka Counties Nevada - Unavailable since 1910, this hard to find publication includes valuable insights into the mining camps of Elko, Lander and Eureka Counties, Nevada. The publication includes important details into the history of mines and mining in these three Nevada counties. 154 ppgs, 12.99

Ore Deposits of the Bullfrog Quadrangle - Unavailable since 1964 and released as "Geology of Bullfrog Quadrangle and Ore Deposits Related to Bullfrog Hills Caldera, Nye County, Nevada and Inyo County, California". The publication includes important details into the geology of mines in the Bullfrog Quadrangle of Nye County, Nevada and Inyo County, California. 52 ppgs, 9.99

Mining in Eureka County Nevada - Unavailable since 1879, this hard to find publication includes valuable insights into the early mining history off Eureka County, Nevada. The publication includes important details into the early history of the mines of Eureka County, as well as their development, production and how their ores were treated. Also included are details on the 1872 Mining Act, as well as the local rules, regulations and customs of the miners in Eureka County.134 ppgs, 12.99

Colorado Mining Books

Ores of The Leadville Mining District - Unavailable since 1926, this publication was originally compiled by the United States Department of Interior. This volume also includes important insights into the ores and mineralization of the Leadville Mining District in Colorado. Topics include historic ore prospecting methods, local geology, insights into ore veins and stockworks, the local trend and distribution of ore channels, reverse faults, shattered rock above replacement ore bodies, mineral enrichment in oxidized and sulphide zones and more. **8.5" X 11", 66 ppgs, Retail Price: $8.99**

Mining in Colorado - Unavailable since 1926, this publication was originally compiled by the United States Department of Interior. This volume also includes important insights into the mining history of Colorado from its early beginnings in the 1850's right up to the mid 1920's. Not only is Colorado's gold mining heritage included, but also its silver, copper, lead and zinc mining industry. Each mining area is treated separately, detailing the development of Colorado's mines on a county by county basis. **8.5" X 11", 284 ppgs, Retail Price: $19.99**

Gold Mining in Gilpin County Colorado - Unavailable since 1876, this publication was originally compiled by the Register Steam Printing House of Central City, Colorado. A rare glimpse at the gold mining history and early mines of Gilpin County, Colorado from their first discovery in the 1850's up to the "flush years" of the mid 1870's. Of particular interest is the history of the discovery of gold in Gilpin County and details about the men who made those first strikes. Special focus is given to the early gold mines and first mining districts of the area, many of which are not detailed in other books on Colorado's gold mining history. **8.5" X 11", 156 ppgs, Retail Price: $12.99**

Mining in the Gold Brick Mining District of Colorado - Important insights into the history of the Gold Brick Mining District, as well as its local geography and economic geology. Also included are the histories and locations of historic mines in this important Colorado Mining District, including the Cortland, Carter, Raymond, Gold Links, Sacramento, Bassick, Sandy Hook, Chronicle, Grand Prize, Chloride, Granite Mountain, Lucille, Gray Mountain, Hilltop, Maggie Mitchell, Silver Islet, Revenue, Roosevelt, Carbonate King and others. In addition to hardrock mining, are also included are details on gold placer mining in this portion of Colorado. **8.5" X 11", 140 ppgs, Retail Price: $12.99**

Ore Deposits of the London Fault of Colorado - First published in 1941, it has been unavailable since those days and sheds important light on the mines and mineral deposits of the London Fault in Central Colorado's Alma Mining District. This publication sheds important light on the gold veins and lead-silver deposits of the Alma Mining District. Included are geologic details on the London Mine, American Mine, Havigorst Tunnel, Ophir Mine, Mosher Tunnel, London-Butte Mine, Venture Shaft, Hard-To-Beat Mine, Oliver Twist Tunnel, Sacramento Mine, Mudsill Mine, Sherwood Mine, Wagner, Barcoe Tunnel and other mines in this important mining region. 110 ppgs., 10.99**

The Mines of Colorado - First published in 1867, it has been unavailable since those days and sheds important light on Colorado's early mining history. Written shortly after the events took place, this publication sheds important light on the Pike's Peak Gold Rush, the discovery of gold on Ralston Creek and Dry Creek in the 1850's, as well as details on the first wave of miners into Colorado and their trials and tribulations as they crossed the Great Plains. Also included are details on early discoveries of lode gold in the mountainous regions of Colorado, details on the early mines hardrock and placer mines, and much more. It is a veritable treasure trove on Colorado's early mining history and will be of great importance to anyone who is interested in the mining of gold or other minerals in Colorado, as well as those interested in the history of the state. 478 ppgs., 29.99**

The La Plata Mining District of Colorado - Originally titled "Geology and Ore Deposits in the Vicinity of the La Plata District of Colorado" and first published in 1949, it has been unavailable since those days and sheds important light on the mines and mineral deposits of the La Plata Mining District of Colorado. 214 ppgs., 19.99**

Washington Mining Books

The Republic Mining District of Washington - Unavailable since 1910, this important publication was originally published by the Washington Geologic Survey and has been unavailable for a century. Topics include the geology, rock formations and the formation of ore deposits in this important mining area of Washington State. Also included are hard to find details on the geology, history and locations of dozens of mines in the area. Some of the mines featured include the New Republic Mine, Ben Hur, Morning Glory, the South Republic Mine, Quilp, Surprise, Black Tail, Lone Pine, San Poil, Mountain Lion, Tom Thumb, Elcaliph and many others. **8.5" X 11", 94 ppgs, Retail Price: $10.99**

The Myers Creek and Nighthawk Mining Districts of Washington - Unavailable since 1911, this important publication was originally published by the Washington Geologic Survey and has been unavailable for a century. Topics include the geology, rock formations and the formation of ore deposits in these important mining areas of Washington State. Also included are hard to find details on the geology, history and locations of dozens of mines in the area. Some of the mines featured include the Grant Mine, Monterey, Nip and Tuck, Myers Creek, Number Nine, Neutral, Rainbow, Aztec, Crystal Butte, Apex, Butcher Boy, Molson, Mad River, Olentangy, Delate, Kelsey, Golden Chariot, Okanogan, Ohio, Forty-Ninth Parallel, Nighthawk, Favorite, Little Chopaka, Summit, Number One, California, Peerless, Caaba, Prize Group, Ruby, Mountain Sheep, Golden Zone, Rich Bar, Similkameen, Kimberly, Triune, Hiawatha, Trinity, Hornsilver, Maquae, Bellevue, Bullfrog, Palmer Lake, Ivanhoe, Copper World and many others. **8.5" X 11", 136 ppgs, Retail Price: $12.99**

The Blewett Mining District of Washington - Unavailable since 1911, this important publication was originally published by the Washington Geologic Survey and has been unavailable for a century. Topics include the geology, rock formations and the formation of ore deposits in this important mining area of Washington State. Also included are hard to find details on the geology, history and locations of dozens of mines in the area. Some of the mines featured include the Washington Meteor, Alta Vista, Pole Pick, Blinn, North Star, Golden Eagle, Tip Top, Wilder, Golden Guinea, Lucky Queen, Blue Bell, Prospect, Homestake, Lone Rock, Johnson, and others. **8.5" X 11", 134 ppgs, Retail Price: $12.99**

Silver Mining In Washington - Unavailable since 1955, this important publication was originally published by the Washington Geologic Survey. Featured are the hard to find locations and details pertaining to Washington's silver mines. **8.5" X 11", 180 ppgs, Retail Price: $15.99**

The Mines of Snohomish County Washington - Unavailable since 1942, this important publication was originally published by the Washington Geologic Survey and has been unavailable for seventy years. Featured are details on a large number of gold, silver, copper, lead and other metallic mineral mines. Included are the locations of each historic mine, along with information on the commodity produced. **8.5" X 11", 98 ppgs, Retail Price: $10.99**

The Mines of Chelan County Washington - Unavailable since 1943, this important publication was originally published by the Washington Geologic Survey and has been unavailable for seventy years. Featured are details on a large number of gold, silver, copper, lead and other metallic mineral mines. Included are the locations of each historic mine, along with information on the commodity. **8.5" X 11", 88 ppgs, Retail Price: $9.99**

Metal Mines of Washington - Unavailable since 1921, this important publication was originally published by the Washington Geologic Survey and has been unavailable for nearly ninety years. Widely considered a masterpiece on the Washington Mining Industry, "Metal Mines of Washington" sheds light on the important details of Washington's early mining years. Featured are details on hundreds of gold, silver, copper, lead and other metallic mineral mines. Included are hard to find details on the mineral resources of this state, as well as the locations of historic mines. Lavishly illustrated with maps and historic photos and complete with a glossary to explain any technical terms found in the text, this is one of the most important works on mining in the State of Washington. No prospector or miner should be without it if they are interested in mining in Washington. **8.5" X 11", 396 ppgs, Retail Price: $24.99**

Gem Stones In Washington - Unavailable since 1949, this important publication was originally published by the Washington Geologic Survey and has been unavailable since first published. Included are details on where to find naturally occurring gem stones in the State of Washington, including quartz crystal, amethyst, smoky quartz, milky quartz, agates, bloodstone, carnelian, chert, flint, jasper, onyx, petrified wood, opal, fire opal, hyalite and others. **8.5" X 11", 54 ppgs, Retail Price: $8.99**

The Covada Mining District of Washington - Unavailable since 1913, this important publication was originally published by the Washington Geologic Survey and has been unavailable for a century. Topics include the geology, rock formations and the formation of ore deposits in this important mining area of Washington State. Also included are hard to find details on the geology, history and locations of dozens of mines in the area. Some of the mines featured include the Admiral, Advance, Algonkian, Big Bug, Big Chief, Big Joker, Black Hawk, Black Tail, Black Thorn, Captain, Cherokee Strip, Colorado, Dan Patch, Dead Shot, Etta, Good Ore, Greasy Run, Great Scott, Idora, IXL, Jay Bird, Kentucky Bell, King Solomon, Laurel, Laura S, Little Jay, Meteor, Neglected, Northern Light, Old Nell, Plymouth Rock, Polaris, Quandary, Reserve, Shoo Fly, Silver Plume, Three Pines, Vernie, White Rose and dozens of others. **8.5" X 11", 114 ppgs, Retail Price: $10.99**

The Index Mining District of Washington - Unavailable since 1912, this important publication was originally published by the Washington Geologic Survey and has been unavailable for a century. Topics include the geology, rock formations and the formation of ore deposits in this important mining area of Washington State. Also included are hard to find details on the geology, history and locations of dozens of mines in the area. Some of the mines featured include the Sunset, Non-Pareil, Ethel Consolidated, Kittaning, Merchant, Homestead, Co-operative, Lost Creek, Uncle Sam, Calumet, Florence-Rae, Bitter Creek, Index Peacock, Gunn Peak, Helena, North Star, Buckeye. Copper Bell, Red Cross and others. **8.5″ X 11″, 114 ppgs, Retail Price: $11.99**

Mining & Mineral Resources of Stevens County Washington - Unavailable since 1920, this important publication was originally published by the Washington Geologic Survey and has been unavailable for a century. Topics include the geology, rock formations and the formation of ore deposits in these important mining areas of Washington State. Also included are hard to find details on the geology, history and locations of hundreds of mines in the area. **8.5″ X 11″, 372 ppgs, Retail Price: $24.99**

The Mines and Geology of the Loomis Quadrangle Okanogan County, Washington - Unavailable since 1972, this important publication was originally published by the Washington Geologic Survey and has been unavailable for a century. Topics include the geology, rock formations and the formation of ore deposits in this important mining area of Washington State. Also included are hard to find details on the geology, history and locations of dozens of gold, copper, silver and other mines in the area. **8.5″ X 11″, 150 ppgs, Retail Price: $12.99**

The Conconully Mining District of Okanogan County Washington - Unavailable since 1973, this important publication was originally published by the Washington Geologic Survey and has been unavailable for a century. Topics include the geology, rock formations and the formation of ore deposits in this important mining area of Washington State, which also includes Salmon Creek, Blue Lake and Galena. Also included are hard to find details on the geology, mining history and locations of dozens of mines in the area. Some of the mines include Arlington, Fourth of July, Sonny Boy, First Thought, Last Chance, War Eagle-Peacock, Wheeler, Mohawk, Lone Star, Woo Loo Moo Loo, Keystone, Hughes, Plant-Callahan, Johnny Boy, Leuena, Gubser, John Arthur, Tough Nut, Homestake, Key and many others **8.5″ X 11″, 68 ppgs, Retail Price: $8.99**

Wyoming Mining Books

Mining in the Laramie Basin of Wyoming - Unavailable since 1909, this publication was originally compiled by the United States Department of Interior. Also included are insights into the mineralization and other characteristics of this important mining region, especially in regards to coal, limestone, gypsum, bentonite clay, cement, sand, clay and copper. **8.5″ X 11″, 104 ppgs, Retail Price: $11.99**

New Mexico Mining Books

The Mogollon Mining District of New Mexico - Unavailable since 1927, this important publication was originally published by the US Department of Interior and has been unavailable for 80 years. Topics include the geology, rock formations and the formation of ore deposits in this important mining area in New Mexico. Of particular focus is information on the history and production of the ore deposits in this area, their form and structure, vein filling, their paragenesis, origins and ore shoots, as well as oxidation and supergene enrichment. Also included are hard to find details, including the descriptions and locations of numerous gold, silver and other types of mines, including the Eureka, Pacific, South Alpine, Great Western, Enterprise, Buffalo, Mountain View, Floride, Gold Dust, Last Chance, Deadwood, Confidence, Maud S., Deep Down, Little Fanney, Trilby, Johnson, Alberta, Comet, Golden Eagle, Cooney, Queen, the Iron Crown, Eberle, Clifton, Andrew Jackson mine, Mascot and others. **8.5″ X 11″, 144 ppgs, Retail Price: $12.99**

The Percha Mining District of Kingston New Mexico - Unavailable since 1883, this important publication was originally published by the Kingston Tribune and has been unavailable for over one hundred and thirty five years. Having been written during the earliest years of gold and silver mining in the Percha Mining District, unlike other books on the subject, this work offers the unique perspective of having actually been written while the early mining history of this area was still being made. In fact, the work was written so early in the development of this area that many of the notable mines in the Percha District were less than a few years old and were still being operated by their original discoverers with the same enthusiasm as when they were first located. Included are hard to find details on the very earliest gold and silver mines of this important mining district near Kingston in Sierra County, New Mexico. **8.5″ X 11″, 68 ppgs, Retail Price: $9.99**

East Coast Mining Books

The Gold Fields of the Southern Appalachians - Unavailable since 1895, this important publication was originally published by the US Department of Interior and has been unavailable for nearly 120 years. Topics include the geology, rock formations and the formation of ore deposits in this important mining area of the American South. Of particular focus is information on the history and statistics of the ore deposits in this area, their form and structure and veins. Also included are details on the placer gold deposits of the region. The gold fields of the Georgian Belt, Carolinian Belt and the South Mountain Mining District of North Carolina are all treated in descriptive detail. Included are hard to find details, including the descriptions and locations of numerous gold mines in Georgia, North Carolina and elsewhere in the American South. Also included are details on the gold belts of the British Maritime Provinces and the Green Mountains. **8.5" X 11", 104 ppgs, Retail Price: $9.99**

Gold Rush Tales Series

Millions in Siskiyou County Gold - In this first volume of the "Gold Rush Tales" series, leading mining historian and editor Kerby Jackson, introduces us to the story of how millions of dollars worth of gold was discovered in Siskiyou County during the California Gold Rush. Lavishly illustrated with photos from the 19th Century, this hard to find information was first published in 1897 and sheds important light onto the gold rush era in Siskiyou County, California and the experiences of the men who dug for the gold and actually found it. **8.5" X 11", 82 ppgs, Retail Price: $9.99**

The California Rand in the Days of '49 - In this second volume of the "Gold Rush Tales" series, leading mining historian and editor Kerby Jackson, introduces us to four tales from the California Gold Rush. Lavishly illustrated with photos from the 19th Century, this hard to find information was first published in 1890's and includes the stories of "California's Rand", details about Chinese miners, how one early miner named Baker struck it rich and also the story of Alphonzo Bowers, who invented the first hydraulic gold dredge. **8.5" X 11", 54 ppgs, Retail Price: $9.99**

More Mining Books

Prospecting and Developing A Small Mine - Topics covered include the classification of varying ores, how to take a proper ore sample, the proper reduction of ore samples, alluvial sampling, how to understand geology as it is applied to prospecting and mining, prospecting procedures, methods of ore treatment, the application of drilling and blasting in a small mine and other topics that the small scale miner will find of benefit. **8.5" X 11", 112 ppgs, Retail Price: $11.99**

Timbering For Small Underground Mines - Topics covered include the selection of caps and posts, the treatment of mine timbers, how to install mine timbers, repairing damaged timbers, use of drift supports, headboards, squeeze sets, ore chute construction, mine cribbing, square set timbering methods, the use of steel and concrete sets and other topics that the small underground miner will find of benefit. This volume also includes twenty eight illustrations depicting the proper construction of mine timbering and support systems that greatly enhance the practical usability of the information contained in this small book. **8.5" X 11", 88 ppgs. Retail Price: $10.99**

Timbering and Mining - A classic mining publication on Hard Rock Mining by W.H. Storms. Unavailable since 1909, this rare publication provides an in depth look at American methods of underground mine timbering and mining methods. Topics include the selection and preservation of mine timbers, drifting and drift sets, driving in running ground, structural steel in mine workings, timbering drifts in gravel mines, timbering methods for driving shafts, positioning drill holes in shafts, timbering stations at shafts, drainage, mining large ore bodies by means of open cuts or by the "Glory Hole" system, stoping out ore in flat or low lying veins, use of the "Caving System", stoping in swelling ground, how to stope out large ore bodies, Square Set timbering on the Comstock and its modifications by California miners, the construction of ore chutes, stoping ore bodies by use of the "Block System", how to work dangerous ground, information on the "Delprat System" of stoping without mine timbers, construction and use of headframes and much more. This volume provides a reference into not only practical methods of mining and timbering that may be employed in narrow vein mining by small miners today, but also rare insights into how mines were being worked at the turn of the 19th Century. **8.5" X 11", 288 ppgs. Retail Price: $24.99**

A Study of Ore Deposits For The Practical Miner - Mining historian Kerby Jackson introduces us to a classic mining publication on ore deposits by J.P. Wallace. First published in 1908, it has been unavailable for over a century. Included are important insights into the properties of minerals and their identification, on the occurrence and origin of gold, on gold alloys, insights into gold bearing sulfides such as pyrites and arsenopyrites, on gold bearing vanadium, gold and silver tellurides, lead and mercury tellurides, on silver ores, platinum and iridium, mercury ores, copper ores, lead ores, zinc ores, iron ores, chromium ores, manganese ores, nickel ores, tin ores, tungsten ores and others. Also included are facts regarding rock forming minerals, their composition and occurrences, on igneous, sedimentary, metamorphic and intrusive rocks, as well as how they are geologically disturbed by dikes, flows and faults, as well as the effects of these geologic actions and why they are important to the miner. Written specifically with the common miner and prospector in mind, the book will help to unlock the earth's hidden wealth for you and is written in a simple and concise language that anyone can understand. **8.5" X 11", 366 ppgs. Retail Price: $24.99**

Mine Drainage - Unavailable since 1896, this rare publication provides an in depth look at American methods of underground mine drainage and mining pump systems. This volume provides a reference into not only practical methods of mining drainage that may be employed in narrow vein mining by small miners today, but also rare insights into how mines were being worked at the turn of the 19th Century. **8.5" X 11", 218 ppgs. Retail Price: $24.99**

Fire Assaying Gold, Silver and Lead Ores - Unavailable since 1907, this important publication was originally published by the Mining and Scientific Press and was designed to introduce miners and prospectors of gold, silver and lead to the art of fire assaying. Topics include the fire assaying of ores and products containing gold, silver and lead; the sampling and preparation of ore for an assay; care of the assay office, assay furnaces; crucibles and scorifiers; assay balances; metallic ores; scorification assays; cupelling; parting' crucible assays, the roasting of ores and more. This classic provides a time honored method of assaying put forward in a clear, concise and easy to understand language that will make it a benefit to even beginners. **8.5" X 11", 96 ppgs. Retail Price: $11.99**

Methods of Mine Timbering - Originally published in 1896, this important publication on mining engineering has not been available for nearly a century. Included are rare insights into historical methods of timbering structural support that were used in underground metal mines during the California that still have a practical application for the small scale hardrock miner of today. **8.5" X 11", 94 ppgs. Retail Price: $10.99**

The Enrichment of Copper Sulfide Ores - First published in 1913, it has been unavailable for over a century. Topics include the definition and types of ore enrichment, the oxidation of copper ores, the precipitation of metallic sulfides. Also included are the results of dozens of lab experiments pertaining to the enrichment of sulfide ores that will be of interest to the practical hard rock mine operator in his efforts to release the metallic bounty from his mine's ore. **8.5" X 11", 92 ppgs. Retail Price: $9.99**

A Study of Magmatic Sulfide Ores - Unavailable since 1914, this rare publication provides an in depth look at magmatic sulfide ores. Some of the topics included are the definition and classification of magmatic ores, descriptions of some magmatic sulfide ore deposits known at the time of publication including copper and nickel bearing pyrrohitic ore bodies, chalcopyrite-bornite deposits, pyritic deposits, magnetite-ileminite deposits, chromite deposits and magmatic iron ore deposits. Also included are details on how to recognize these types of ore deposits while prospecting for valuable hardrock minerals. **8.5" X 11", 138 ppgs. Retail Price: $11.99**

The Cyanide Process of Gold Recovery - Unavailable since 1894 and released under the name "The Cyanide Process: Its Practical Application and Economical Results", this rare publication provides an in depth look at the early use of cyanide leaching for gold recovery from hardrock mine ores. This volume provides a reference into the early development and use of cyanide leaching to recover gold. **8.5" X 11", 162 ppgs. Retail Price: $14.99**

California Gold Milling Practices - Unavailable since 1895 and released under the name "California Gold Practices", this rare publication provides an in depth look at early methods of milling used to reduce gold ores in California during the late 19th century. This volume provides a reference into the early development and use of milling equipment during the earliest years of the California Gold Rush up to the age of the Industrial Revolution. Much of the information still applies today and will be of use to small scale miners engaging in hardrock mining. **8.5" X 11", 104 ppgs. Retail Price: $10.99**

Leaching Gold and Silver Ores With The Plattner and Kiss Processes - Mining historian Kerby Jackson introduces us to a classic mining publication on the evaluation and examination of mines and prospects by C.H. Aaron. First published in 1881, it has been unavailable for over a century and sheds important light on the leaching of gold and silver ores with the Plattner and Kiss processes. **8.5" X 11", 204 ppgs. Retail Price: $15.99**

The Metallurgy of Lead and the Desilverization of Base Bullion - First published in 1896, it has been unavailable for over a century and sheds important light on the the recovery of silver from lead based ores. Some of the topics include the properties of lead and some of its compounds, lead ores such as galenite, anglesite, cerussite and others, the distribution of lead ores throughout the United States and the sampling and assaying of lead ores. Also covered is the metallurgical treatment of lead ores, as well as the desilverization of lead by the Pattinson Process and the Parkes Process. Hofman's text has long been considered one of the most important early works on the recovery of silver from lead based ores. 8.5" X 11", 452 ppgs. **Retail Price: $29.99**

Ore Sampling For Small Scale Miners - First published in 1916, it has been unavailable for over a century and sheds important light on historic methods of ore sampling in hardrock mines. Topics include how to take correct ore samples and the conditions that affect sampling, such as their subdivision and uniformity. Particular detail is given to methods of hand sampling ore bodies by grab sample, pipe sample and coning, as well as sampling by mechanical methods. Also given are insights into the screening, drying and grinding processes to achieve the most consistent sample results and much more. 8.5" X 11", 124 ppgs. **Retail Price: $12.99**

The Extraction of Silver, Copper and Tin from Ores - First published in 1896, it has been unavailable for over a century and sheds important light on how historic miners recovered silver, copper and tin from their mining operations. The book is split into three sections, including a discussion on the Lixiviation of Silver Ores, the mining and treatment of copper ores as practiced at Tharsis, Spain and the smelting of tin as it was practiced by metallurgists at Pulo Brani, Singapore. Also included is an overview and analysis of these historic metal recovery methods that will be of benefit to those interested in the extraction of silver, copper and tin from small mines. 8.5" X 11", 118 ppgs. **Retail Price: $14.99**

The Roasting of Gold and Silver Ores - First published in 1880, it has been unavailable for over a century and sheds important light on how historic miners recovered gold and silver rom their mining operations. Topics include details on the most important silver and free milling gold ores, methods of desulphurization of ores, methods of deoxidation, the chlorination of ores, methods and details on roasting gold and silver ores, notes on furnaces and more. Also included are details on numerous methods of gold and silver recovery, including the Ottokar Hofman's Process, the Patera Process, Kiss Process, Augustin Process, Ziervogel Process and others. 8.5" X 11", 178 ppgs. **Retail Price: $19.99**

The Examination of Mines and Prospects - First published in 1912, it has been unavailable for over a century and sheds important light on how to examine and evaluate hardrock mines, prospects and lode mining claims. Sections include Mining Examinations, Structural Geology, Structural Features of Ore Deposits, Primary Ores and their Distribution, Types of Primary Ore Deposits, Primary Ore Shoots, The Primary Alteration of Wall Rocks, Alterations by Surface Agencies, Residual Ores and their Distribution, Secondary Ores and Ore Shoots and Vein Outcrops. This hard to find information is a must for those who are interested in owning a mine or who already own a lode mining claim and wish to succeed at quartz mining. 8.5" X 11", 250 ppgs. **Retail Price: $19.99**

Garnets: Their Mining, Milling and Utilization - First published in 1925, it has been unavailable since those days and sheds important light on the mining, milling and utilization of garnets. Included are details on the characteristics of garnets, where they are found and how they were mined. 78 ppgs, 10.99

Gemstones and Precious Stones of North America - Leading mining historian Kerby Jackson introduces us to a classic mining publication on the gems and precious stones of the United States, Canada and mexico. First published in 1890, it has been unavailable since those days and sheds important light on the gems and precious stones that may be found in North America. Included are chapters on diamonds, corundum, sapphire, ruby, topaz, emerald, disapore, spinel, turquoise, tourmaline, garnets, beyrl, peridot, zircon, quartz crystals, feldspars, pearls and many others. Included are details on where these gems and precious stones may be found throughout North America, as well as their characteristics. 360 ppgs, 24.99

Mining Camps and Mining Districts - First released in 1885 by Charles Howard Shinn under the title "Mining Camps: A Study in American Frontier Government", this publication offers a unique look at how early gold miners established their own forms of representative government during the California Gold Rush. Drawing on the the early mining codes of mideviel German miners in the Harz Mountains, on the mining customs of the Cornish tin miners and early Spanish mining laws introduced into California, the miners established the first governments in the American West. 340 ppgs, 24.99

BLM Field Handbook for Mineral Examiners - Leading mining historian Kerby Jackson introduces us to a classic mining publication on mine evaluation. First published in 1962, this work sheds important light on the techniques of BLM Mineral Examiners to perform validity on mining claims. 132 ppgs, 10.99

<u>Six Months In The Gold Mines During The California Gold Rush</u> - Unavailable since 1850, this important work is a first hand account of one "49'ers" personal experience during the great California Gold Rush, shedding important light on one of the most exciting periods in the history of not only California, but also the world. Compiled from journals written between 1847 and 1849 by E. Gould Buffum, a native of New York, "Six Months In The Gold Mines During The California Gold Rush" offers a rare look into the day to day lives of the people who came to California to work in her gold mines when the state was still a great frontier. **8.5" X 11", 290 ppgs. Retail Price: $19.99**

<u>The Discovery of Gold in Australia</u> - First published in 1852, it has been unavailable since those days and sheds important light on Australia's gold mining history. Included are rare communications between British agents and the British Crown when gold was first discovered in Australia in 1851. This rare text contains hard to find details on Australia's first mining camps and Britain's early attempts to provide for the orderly regulation of gold mines in that part of the world. Also of interest are hard to find extracts of articles that appeared in the early colonial newspapers that did their best to report on Australia's gold rush as it took place.
102 ppgs, 10.99

www.ingramcontent.com/pod-product-compliance
Lightning Source LLC
Chambersburg PA
CBHW080835180526
45168CB00006B/2691